FRESHWATER PHYTOPHARMACEUTICAL COMPOUNDS

FRESHWATER PHYTOPHARMACEUTICAL COMPOUNDS

Santhanam Ramesh
Rajabalaya Rajan
Ramasamy Santhanam

CRC Press
Taylor & Francis Group
Boca Raton London New York

CRC Press is an imprint of the
Taylor & Francis Group, an **informa** business

Reprint 2019
CRC Press
Taylor & Francis Group
6000 Broken Sound Parkway NW, Suite 300
Boca Raton, FL 33487-2742

ISBN:9780367268978

Library of Congress Cataloging-in-Publication Data

Ramesh, Santhanam, 1979-
Freshwater phytopharmaceutical compounds / Santhanam Ramesh, Rajan Rajabalaya, Ramasamy Santhanam.
pages cm
Includes bibliographical references and index.
ISBN:9780367268978
1. Materia medica, Vegetable. 2. Freshwater plants--Therapeutic use. I. Rajabalaya, Rajan. II. Santhanam, Ramasamy, 1946- III. Title.

RS164.R326 2013
615.1--dc23 2013008699

Visit the Taylor & Francis Web site at
http://www.taylorandfrancis.com

and the CRC Press Web site at
http://www.crcpress.com

Printed and bound by Repro India Ltd.

Contents

Preface

Bioprospecting, i.e., identification of potential organisms yielding bioactive compounds, is considered to be the prerequisite in the development of new drugs. Unfortunately, until now, no efforts have been made to bring together both biologists and pharmacists in drug-oriented research programs for the benefit of pharmaceutical industries. Keeping these in view, a foundation has now been laid in the amalgamation of the biology and pharmacy disciplines. This approach, when encouraged and augmented, would certainly help the R&D wings of pharmaceutical industries in the identification and large-scale collection of pharmaceutically important organisms that would be possible only when their common names, classification, global distribution, habitats, and biology are known. Considering these aspects, the present publication has been brought forth. The different chapters deal with species of freshwater plants possessing bioactive compounds for various activities, such as antimicrobial, antiprotozoal, antihelmintic, antiviral, anticancer, antioxidant, anti-inflammatory and analgesic, antidiabetic, anticoagulant, central nervous system (CNS) disorders, hypolipidemic, hepatoprotective, etc.

It is strongly hoped that this publication will be widely welcomed not only by pharmacy and biology disciplines but also by related disciplines, such as biotechnology, life sciences, and environmental sciences. The authors express their sincere thanks to Adv. Dr. Krishnadas, Managing Trustee of Nehru Group of Institutions Thiruvilwamala, Kerala, India, for his kind foreword. We are highly indebted to Prof. Dr. V. Sankar, Head of the Department of Pharmacy Practice, PSG College of Pharmacy, Coimbatore, India, for his valued comments and suggestions. We also thank Mr. C. Rajamanickam of Vijaya Technical Institute, Parangipettai, Tamilnadu, India, for his excellent typing of the manuscript and secretarial assistance, and Mrs. J. Albin Panimalar for help in photography.

Suggestions from the readers are welcome.

About the Authors

Dr. Santhanam Ramesh, Professor, Department of Pharmaceutics, Nehru College of Pharmacy, Thiruvilwamala, Kerala, obtained his PhD degree from Jawaharlal Nehru Technological University, Hyderabad, India, in 2011. His fields of specialization are biopharmaceutical products, pharmaceutical nanotechnology, and novel drug delivery systems.

Dr. Ramesh has seven years of teaching/overseas experience. He has authored two books, titled *Marine Pharmaceutical Compounds* and *A Text Book of Novel Drug Delivery Systems*, and 10 research papers published in nationally and internationally reputed journals. He is presently serving as an associate member of the Academy Pharmacy Group, Royal Pharmaceutical Society, Great Britain, and as a member of American Nano Society.

Dr. Rajabalaya Rajan, lecturer III, International Medical University, Bukit Jalil, Kuala Lumpur, Malaysia, obtained his PhD from Jadavpur University, Kolkata, India, in 2009. His fields of specialization are biopharmaceutical products and transdermal and topical drug delivery. He has 13 years of teaching experience and has written 20 research papers. He is presently a member of the editorial board of the *International Journal of Pharmacy* and is a member of the former Industrial Research and Malaysian Pharmaceutical Society.

Dr. Ramasamy Santhanam is the former Dean of Fisheries College and Research Institute, Tamilnadu Veterinary and Animal Sciences University, India. His fields of specialization are fisheries environment and aquatic biology. He is presently serving as a fisheries expert for various government and nongovernment organizations in India. Dr. Santhanam has published 10 textbooks on various aspects of fisheries science and 70 research papers. He was an active member of American Fisheries Society, United States; World Aquaculture Society, United States; Global Fisheries Ecosystem Management Network (GFEMN), United States; and IUCN's Commission on Ecosystem Management, Switzerland.

1 Introduction

Our ancestors largely relied upon drugs such as antibiotics, pain killers, cancer therapies, and anti-inflammatory drugs derived from organic products (bioactive natural products or secondary metabolites) of plants and animals from time immemorial to improve and enrich their own lives. It has been reported that about 50% of the currently marketed drugs have their origin in natural products yielded by plants and animals. In the case of anti-infective and anticancer agents, the natural products have contributed more than 70%. However, modern medicines and therapies are either not capable of completely controlling the target diseases or are in limited supply. Further, owing to the emergence of antibiotic resistance in bacteria and the vast spread of devastating diseases such as AIDS, there is an urgent need for the discovery of new drugs.

In this context, natural products of freshwater plants provide a diverse and unique source of novel bioactive lead compounds for drug discovery. The freshwater blue-green algae (cyanobacteria) that are the primitive forms of life have been reported to yield several bioactive compounds for the development of antiviral, antitumor, antibacterial, and anti-HIV drugs.

Freshwater plants (algae and macrophytes) inhabiting stationary water bodies (lacustrine systems) such as lakes, ponds, and reservoirs and flowing waters (streams and rivers) are important in the health of the ecosystems, as they help the water bodies largely by enhancing dissolved oxygen content, nutrient cycling, and substrate stabilization, and reducing or regulating water temperature, unwanted phytoplankton blooms, and hazardous chemicals. Besides improving water quality, these plants also provide refuge or habitat and food for associated micro- and macrofauna.

FRESHWATER PLANTS AS SOURCES OF DRUGS

Herbal medicines have been in use as prophylactic measures and for curative treatments since the dawn of human civilization. Crude extracts of herbal origin are nowadays widely used in pharmaceutical formulations, and the search for drugs (Sivarajan and Balachandran, 1994; Li, 2011) and dietary supplements from freshwater has accelerated in recent years.

FRESHWATER PLANTS AND THEIR BIOACTIVE COMPOUNDS

Freshwater plants have been reported to produce a variety of structurally novel bioactive compounds or secondary metabolites such as antibiotics, mycotoxins, alkaloids, and phenolic compounds, which are considered to be valuable sources of pharmaceutical or biomedical compounds in the discovery of modern drugs.

TYPES OF BIOACTIVE COMPOUNDS FROM FRESHWATER PLANTS AND THEIR ACTIVITIES

The bioactive compounds produced by different freshwater plants include glycosides, glucosinolates, saponins, anthroquinone glycosides, flavonoids, proanthocyanidins, tannins, mono- and sesquiterpenoids, phenyl propanoids, diterpenoids, tropane alkaloids, pyrrolizadine alkaloids, isoquinoline alkaloids, methylxanthine alkaloids, pseudoalkaloids, furocoumarines, naphthodianthrones, and proteins and peptides (Wangchuk, 2004). The broad activities of these bioactive compounds vary from antimicrobial and antiviral activity to cytotoxicity and immunomodulatory activity. The specific activities of different bioactive compounds produced by freshwater plants are given below:

1. Antimicrobial (antibacterial, antimycobacterial-antitubercular, and antifungal)
2. Antiprotozoal (antimalarial-antiplasmodial, antitrypanosomal, and antileishmanial)
3. Antihelmentic (nematocidal)
4. Antiviral (antiretroviral and antiherpetic)
5. Anticancer (antitumor, antiproliferative, antileukemic, antineoplasmic, antimitotic, and cytotoxic)
6. Antioxidant
7. Anti-inflammatory and analgesic
8. Antidiabetic, antidiuretic, diuretic, and uretic
9. Blood-related activities such as anticoagulant (anticlotting or antithrombotic), antiplatelet, platelet aggregation, hematopoietic, granulopoietic and hypocholesterolemic (hypolipidemic)
10. Central nervous system (CNS)-related disorders such as antiepileptic (anticonvulsant), anxiolytic (antianxiety, tranquilizing, and sedative), and antiparkinsonian
11. Miscellaneous activities such as antihepatitis (hepatoprotective), antipyretic, antiulcerogenic, protease inhibitor, etc.

Though considerable progress has been made in the isolation of bioactive compounds from freshwater plants, constraints such as collection of pharmaceutical freshwater plants and their identification (freshwater bioprosecting) need to be mitigated through concerted efforts and advancements in research.

2 Antimicrobial Compounds

Antimicrobial agents produced from freshwater plants are broadly categorized into antibacterial, antifungal, and antimycobacterial agents. Such compounds include phenolics, terpenoids, alkaloids, lectins, polypeptides, and polyacetylenes. Potential antituberculous agents isolated from plants include berberine, lichoisoflavone, erygibisoflavone, phaseollidin, erythrabyssin II, and triptanthrin.

PITHOPHORA OEDOGONIA (MONTAGNE) WITTROCK, 1877

Phylum: Chlorophyta
Class: Chlorophyceae
Order: Cladophorales
Family: Cladophoraceae
Distribution: Tropical and temperate regions of the world; Europe, Asia, Australia, New Zealand
Habitat: Freshwater lakes and ponds, marine habitats
Description: This species has branched, free-floating solitary or paired filaments that form mats. The thallus has large terminal and intercalary akinetes. The cells of the filaments are cylindrical and longer at the ends of the branches. The asexual reproduction of this species is by fragmentation and akinetes.
Compounds and activities: The methanol extracts of the dried material of this species possess phenol, saponin, tannin, amino acid, steroid, and flavonoid, which have antibacterial activities toward Gram-positive bacteria, viz., *Bacillus subtilis* and *Staphylococcus aureus*, and Gram-negative bacteria, viz., *Escherichia coli* and *Salmonella typhimurium*. The growth inhibiton zone diameters (mm) in these bacterial species were found to be 18.1 ± 0.44, 17.1 ± 0.54, 15.80 ± 0.53, and 12.64 ± 0.42, respectively (Singh and Chaudhary, 2010).

SPIROGYRA DECIMINA (O.F. MULLER) DEMORTIER, 1822

Phylum: Chlorophyta
Class: Chlorophyceae
Order: Zygnematales
Family: Zygnemataceae
Distribution: Europe, America, Asia, Australia, New Zealand
Habitat: Freshwater habitats
Description: This unbranched filamentous species is several centimeters long and 100 μm wide. The chloroplasts, which are ribbon shaped, scalloped, or serrated, are helically arranged in the filaments. The cells are connected end to end.
Compounds and activities: The bioactive compounds isolated from this species are unsaturated fatty acids, organic acids, and phenolic compounds. The methanol extract of this species at a concentration of 0.20 mg/ml showed

antibacterial activity against *S. aureus* (inhibition zone diameter, 12 mm) and *P. mirabilis* (9 mm). Similarly, its hexane extract (0.37 mg/ml) showed activity against *S. aureus* (9 mm) (Prakash et al., 2011).

CHLORELLA VULGARIS BEYENINCK (BEIJERINCK), 1890

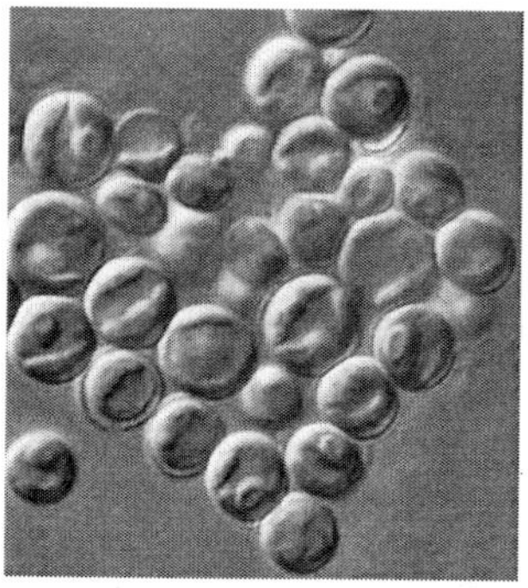

Phylum: Chlorophyta Class: Trebouxiophyceae
Order: Chlorellales Family: Chlorellaceae
Distribution: Ireland, Europe, South America, Southwest Asia, Central Asia, Southeast Asia, Australia, New Zealand
Habitat: Freshwater habitats
Description: This is a single-celled, free-living species. The cells (3–10 μm diameter) are spherical without flagella. The green photosynthetic pigments chlorophyll *a* and *b* are present.
Compounds and activities: The bioactive compounds isolated from this species include 2-butanol, 1,1-diethoxybutanol, 1-hexyl-2-nitrocyclohexane, heptanol, octanol, and 3-methyl-(S)-ethane. Among these compounds, 3-methyl-(S)-ethane showed maximum activity (90.8%) (Al-Wathnani et al., 2012). Other compounds, such as aminopeptides (glutathiones), acidic glycoproteins, galactose-rich carbohydrates, nucleic acids, bromoacetic acid, and pentadecyl ester, have also been isolated from this species. All these compounds have shown antibacterial, antiviral, antitumor, antimetastatic, and immunopotentiating effects. The methanolic and hexanolic extracts of this species have shown antibacterial properties against *S. aureus*, *Staphylococcus epidermidis*, *B. subtilis*, *E. coli*, and *Salmonella typhi*. The flavonoids isolated from this species, however, have shown antioxidative activities. This species also has a chemopreventive effect by inducing apoptosis by decreasing the expression of Bcl-2 and increasing the expression of caspase 8 in hepatocarcinogenesis-induced rats (Azamai et al., 2009). Oral administration of hydrolysates of this species protects the human body from heavy metal pollution. As a food supplement, this species has been found to help the human body improve liver function, lower blood pressure, boost immune function, and reduce obesity.

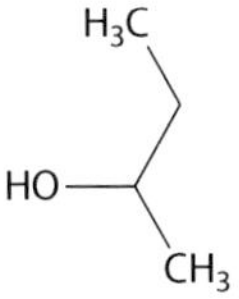

2-butanol

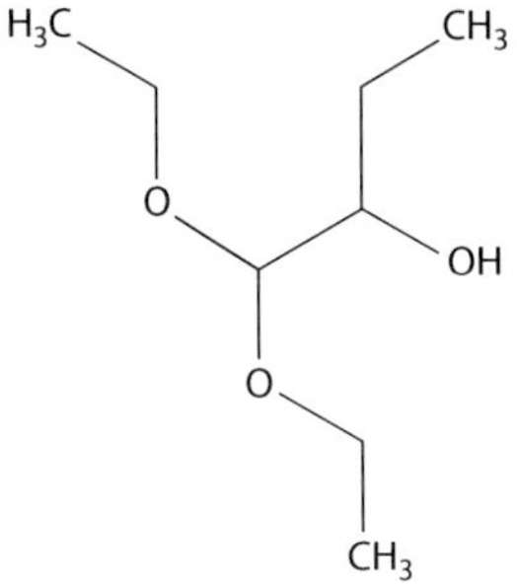

1,1,-diethoxy butanol

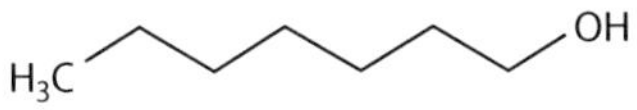

heptanol

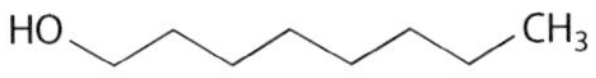

octanol

HAEMATOCOCCUS FLUVIALIS (FLOWTOW, 1844)

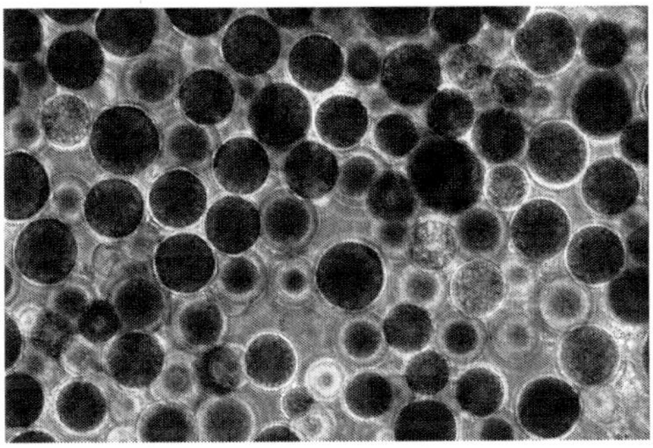

Phylum: Chlorophyta Class: Chlorophyceae
Order: Volvocales Family: Haematococcaceae
Distribution: Temperate regions of the world: Europe, South America, Asia, Pacific Islands
Habitat: Ponds, rainwater pools, coastal rock pools
Description: This microscopic species has unicellular, flagellated, and motile green cells.
Compounds and activities: The short-chain fatty acids of this species show antimicrobial properties (Meizoso et al., 2010). Its vitamin E and simple phenols possess antioxidant activities. The extracts of this species show powerful anti-inflammatory activities. The astaxanthin and cantaxanthin of this species offer vast scope in applications relating to pharmaceuticals and cosmetics.

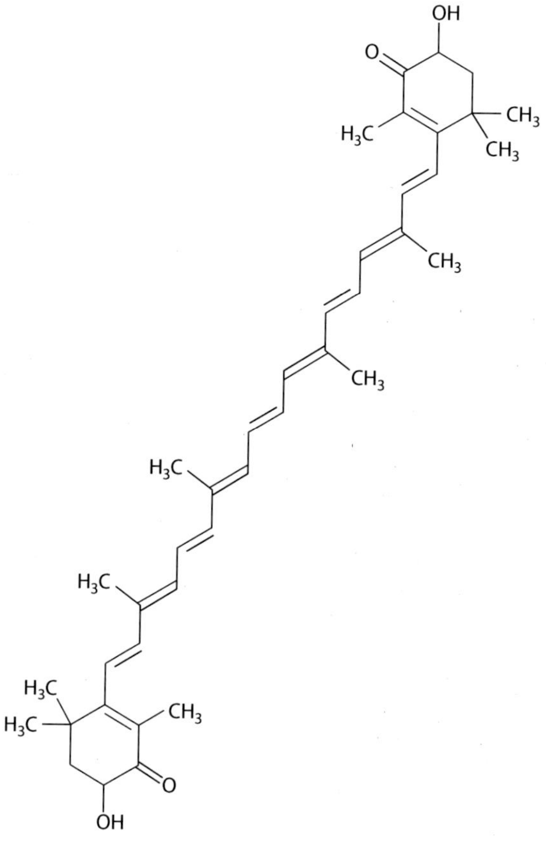

astaxanthin

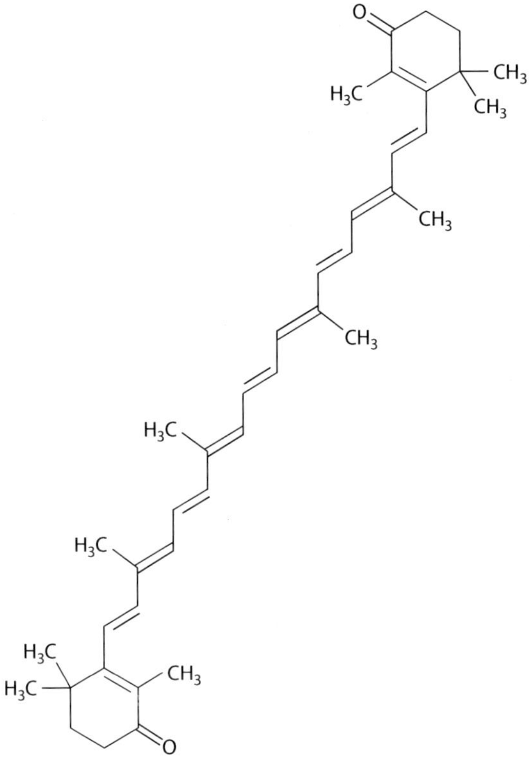

cantaxanthine

SYNECHOCOCCUS LEOPOLIENSIS (RACIBORSKI) KOMREK

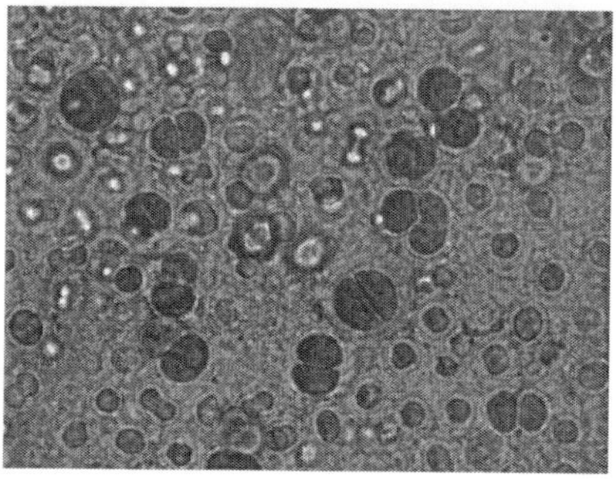

Phylum: Cyanobacteria
Class: Cyanophyceae
Order: Synechococcales
Family: Synechococcaceae
Distribution: Europe
Habitat: Freshwaters to hyposaline areas
Description: The cells of this unicellular species are small and ovoid or cylindrical with a size range of 0.5–2.0 μm. This species has chlorophyll *a* as the main pigment and phycobiliproteins as the accessory pigments.
Compounds and activities: The extracts of planktonic freshwater cyanobacteria are known for their antibacterial properties (Ostensvik et al., 2011). The extracts of this species possess antibacterial activity and act against Gram-positive *Staphylococcus aureus*. Very high antibacterial activity was detected in this species by using the carbon source galactose in combination with the nitrogen source alanine or by using arabinose with methionine. Further, maximum growth and antibacterial activity of this species were noted at 35°C and pH 8 (Noaman et al., 2004).

SPIRULINA PLATENSIS (GOMONT) GEITLER

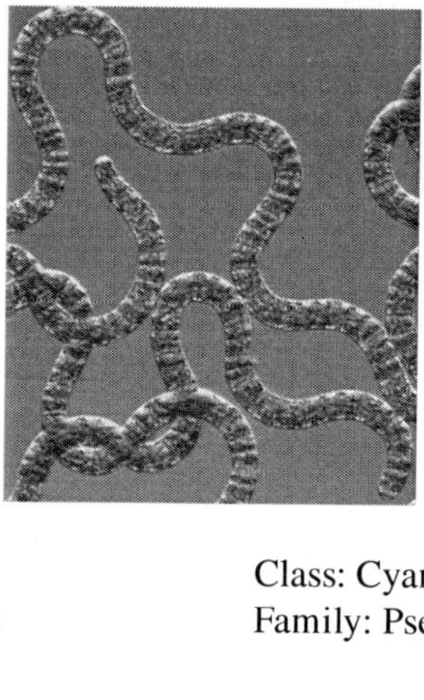

Phylum: Cyanobacteria
Class: Cyanophyceae
Order: Pseudanabaenales
Family: Pseudanabaenaceae

Distribution: Central and South America, Africa, Europe, Southwest Asia

Habitat: Lakes and warm alkaline brackish waters

Description: This species has microscopic, free-floating filaments that are dark green and spirally coiled. Its photosynthetic pigments include chlorophylls, carotenoids, and phycocyanin. This species is known for its high protein content (70% by dry weight) of high nutritional values. Vitamin B12 and trace minerals such as iron, calcium, magnesium, manganese, potassium, selenium, zinc, and bioflavonoids are also present in this species.

Compounds and activities: The bioactive compounds of this species include saturated fatty acids; unsaturated fatty acids, like oleic and linoleic acids; and unsaponiafiable compounds, such as cholesterol, campesterol, stigmasterol, β-sitosterol, and nitrocyclohexane-2-hexyl-l, which are known for their antimicrobial, antioxidative, antiviral, and immunostimulating activities. Among the different compounds, nitrocyclohexane-2-hexyl-1 showed a maximum activity of 92.1% (Al-Wathnani et al., 2012). The methanol extracts of this species have shown the presence of a bioactive compound, viz., quercetin, which has antibacterial activities against pathogenic bacteria such as *E. coli*, *Salmonella typhimurium*, and *Streptococcus faecalis*. The immune-stimulating and antiviral properties are achieved by activating microphages, NK cells, T cells, and B cells and stimulating the production of interferon γ and other cytokines. The natural substances present in this species are known to inhibit several viruses by blocking viral absorption, penetration, and replication. The phycocyanin content of this species has been reported to act against herpes simplex virus type 2, besides inhibiting the growth of human leukemia k562 cells. Further, this compound influences serum cholesterol concentration and imparts a potent hypocholesterolemic activity.

LIMNOTHRIX REDEKEI (VAN. GOOR) M.E. MEFFERT (= *OSCILLATORIA REDEKEI*)

Phylum: Cyanobacteria | Class: Cyanophyceae

Order: Pseudanabaenales | Family: Pseudanabaenaceae

Distribution: Europe, North America

Habitat: Freshwater habitats

Description: This free-floating and filamentous species has solitary filaments that are straight, slightly curved, or irregularly screw-like coiled. These many-celled filaments are isopolar and without any sheath or with a fine colorless facultative sheath. The trichome is cylindrical and is composed of cells (2–6 μm wide) that do not taper at the ends. However, cylindrical apical cells are usually present in the trichome. The cells are yellowish or pinkish and produce facultative gas vesicles. The heterocysts and akinetes are, however, absent.

Compounds and activities: The bioactive compounds isolated from this species include α-dimorphecolic acid (9-hydroxy-IOE), 12 Z-octadecadienoic acid (9-HODE), and coriolic acid, a 13-hydroxy-9Z,11E-octadecadienoic acid (13-HODE). In agar plate tests, these unsaturated hydroxyl fatty acids inhibit the growth of Gram-positive bacteria, *Bacillus subtilis* SBUG 14, *Micrococcus flavus* SBUG, and *Staphylococcus aureus* SBUG 11 and ATCC 25923 (Mundt et al., 2003).

OSCILLATORIA HAMELII FREMY, 1930

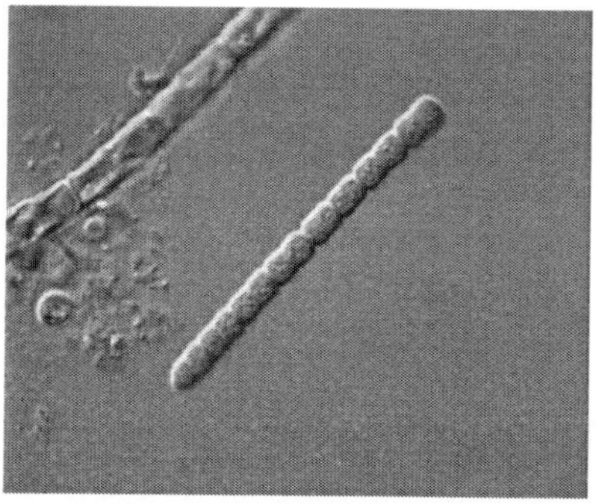

Phylum: Cyanobacteria
Class: Cyanophyceae
Order: Oscillatoriales
Family: Oscillatoriaceae
Distribution: Europe
Habitat: Freshwater habitats
Description: The trichome (8 × 5 μm) of this species is not attenuated toward the ends and is slightly constricted. The granules are absent and the terminal cells are rounded.
Compounds and activities: The diethyl ether extracts of this species have both saturated and unsaturated fatty acids at 83 and 10%, respectively. The other bioactive compounds present in this species include tetramine, spermine, and piperazine derivatives. This species has shown antimicrobial activities against *Escherichia coli*, *Bacillus subtilis*, *Staphylococcus albus*, *Streptococcus faecalis*, and *Candida albicans*, and the values of inhibition zone diameter (mm) were found to be 15, 14, 14, 12, and 12, respectively (Shanab, 2007).

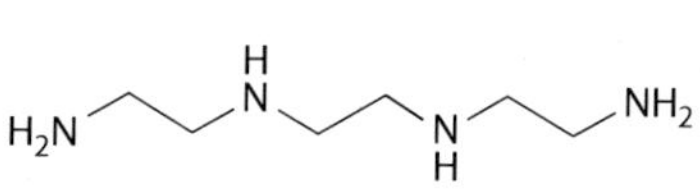

tetramine

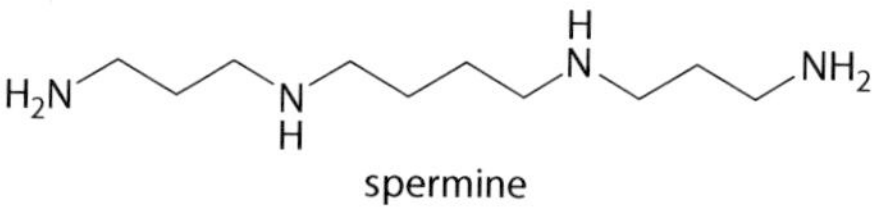

spermine

OSCILLATORIA PLATENSIS (NODST) BOURELLY

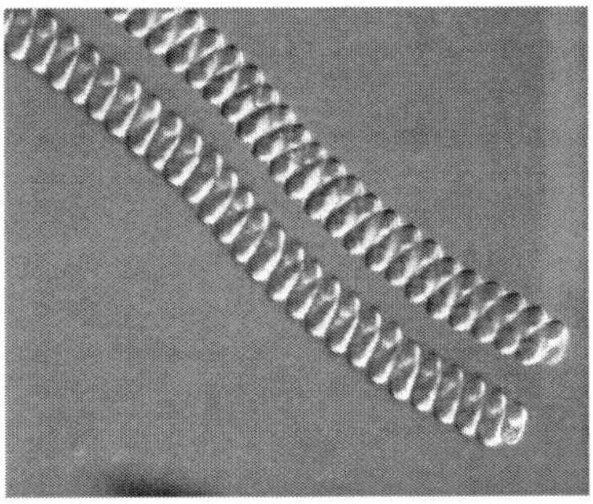

Phylum: Cyanobacteria
Class: Cyanophyceae
Order: Oscillariatoles
Family: Oscillatoriaceae
Distribution: Europe
Habitat: Ponds, stagnant waters
Description: This unbranched filamentous algae form extensive mats at the surface. The filaments in colonies may slide back and forth.
Compounds and activities: The diethyl ether extracts of this species possess both saturated and unsaturated fatty acids at 75 and 17%, respectively. The bioactive compounds present in this species include tetramine, spermine, and piperazine derivatives. This species has shown antimicrobial activities against *E. coli*, *B. subtilis*, *S. aureus*, *S. faecalis*, and *C. albicans*. The values of inhibition zone diameter (mm) for the above species were 13, 11, 16, 11, and 14, respectively (Shanab, 2007).

OSCILLATORIA SANCTA KUTZING EX GOMONT 1892

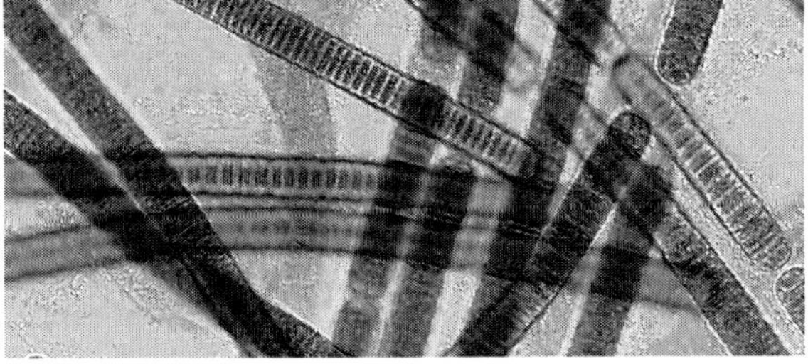

Phylum: Cyanobacteria
Class: Cyanophyceae
Order: Oscillatoriales
Family: Oscillatoriaceae
Distribution: Ireland, Europe, South America, Africa, Southwest Asia, Australia, New Zealand, Pacific Islands, Caribbean Islands
Habitat: Moist rock, associated with *Nostoc linckia*, moist soil, lakes, ponds, shallow pools

Description: This species, which shows olive-green, bright blue, green, or grayish coloration, forms rectangular mucilaginous blocks. The trichomes (3 × 11 µm) are straight or bent and are distinctly constricted at cross walls. The apical cells are flattened, slightly capitates, or wart-like.

Compounds and activities: The extracts of this species have sanctolide A (a 14-membered PK-NRP-hybrid macrolide) (N-methyl enamide and 2-hydroxy isovaleric acid) as the bioactive compound, which possesses antimicrobial properties (Kang et al., 2012).

PLANKTOTHRIX RUBESCENS (DE CANDOLLE EX GOMONT) ANAGNOSTIDIS & KOMAREK, 1988 (= *OSCILLATORIA RUBESCENS*)

Phylum: Cyanobacteria Class: Cyanophyceae

Order: Oscillatoriales Family: Phormidiaceae

Distribution: Southeast Asia, Southwest Asia, Australia, New Zealand, South America

Habitat: Freshwater habitats, lakes

Description: The blooms of this species may show distinctive red color. The cells, which may attach end to end, may form filaments of several millimeters in length. Gas vesicles are found to occupy the whole cell volume. The trichomes (5–10 µm wide) are reddish or pinkish red. Apical cells are narrow with triangular calyptra.

Compounds and activities: The extracts of this species contain saturated and unsaturated fatty acids at 82 and 13%. The active ingredients in the extract are piperazine and spermine, which possess antimicrobial activities. The inhibitory activities shown by this species relate to *E. coli*, *B. subtilis*, *Staphylococcus albus*, *S. faecalis*, and *Candida albicans*. The values of inhibition zone diameter (mm) for these microbial species were 18, 14, 12, 13, and 17, respectively (Shanab, 2007). Further, the bioactive compounds isolated from this species, viz., planktopeptin BL 1125, BL 843, and BL 1061, have been reported to be protease, elastase, and chymotrypsin inhibitors.

PHORMIDIUM AUTUMNALE (C. AGARDH) GOMONT, 1892

Phylum: Cyanobacteria Class: Cyanophyceae
Order: Oscillatoriales Family: Phormidiaceae
Distribution: Australia, New Zealand, Europe, Ireland, Southwest Asia, South America
Habitat: Creeks, streams (subaerophytic habitats)
Description: The filaments of this species are unbranched and are rarely solitary. Though microscopic initially, it becomes macroscopic, of several centimeters in diameter, subsequently. A facultative sheath is present. The trichome (3–15 μm wide) is isopolar, more or less straight, coiled, or waved. It is uniserial and unbranched. The cells of the trichome are cylindrical or barrel shaped and without aerotops.
Compounds and activities: The crude extract (acetone/methanol/di-ethyl-ether, 5:2:1 v/v) of this species has yielded the bioactive compound 1-hexyl-2-nitrocyclohexane, which showed a maximum activity of 91.7% against human pathogenic bacteria. The extract also showed moderate activity against *Bacillus subtilis* and high activity against *S. sonnei* (Al-Wathnani, 2012).

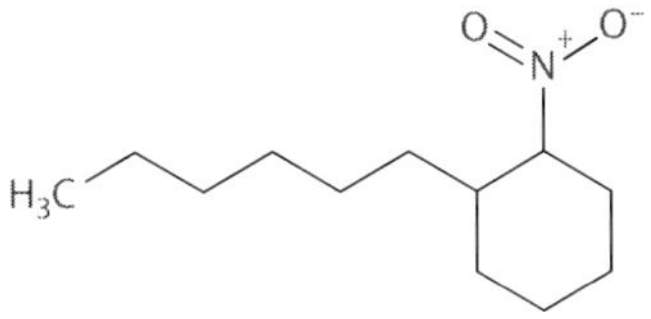

1-hexyl-2-nitrocyclohexane

NOSTOC MINUTUM DESMAZIERES, 1831

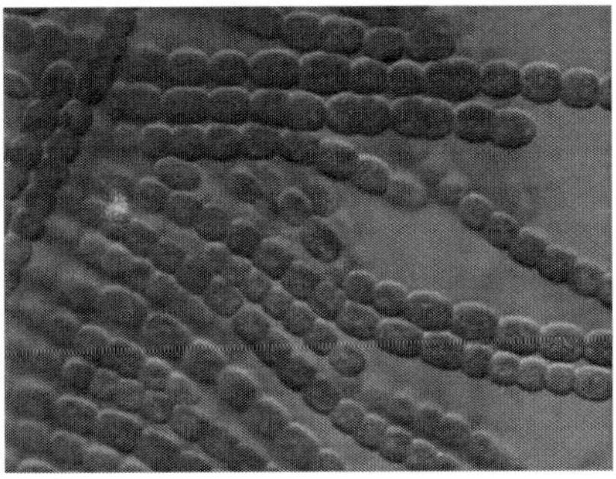

Phylum: Cyanobacteria Class: Cyanophyceae
Order: Nostocales Family: Nostocaceae
Distribution: Europe, Asia, Australia, New Zealand, Japan
Habitat: Freshwater habitats
Description: The colonies of this small macroscopic species are gelatinous and leathery in structure. Each colony consists of many trichomes that are embedded in a communal mucilage.

Compounds and activities: The bioactive compound, muscoride A, isolated from this species has antibacterial properties (Rao, 2004), and nostopeptin A and B (cyclic depsipeptides) of this species are elastase and chymotrypin inhibitors. Other compounds present in this species are microviridins B and H (cyclic peptides). The extract of this species has also shown antibacterial and antifungal activities, and the values of inhibition zone diameter (mm) recorded for the bacterial species, *S. epidermidis*, *S. aureus*, *S. haemolyticus*, *P. vulgaris*, and *S. sonnei*, were found to be 22, 24, 16, 18, and 15, respectively. Similarly, the values for the antifungal species, *C. krusei*, *C. albicans*, *A. niger*, *A. fumigatus*, and *C. neoformans*, were 9, 8, 9, 7, and 10 mm, respectively (Ramamurthy et al., 2012).

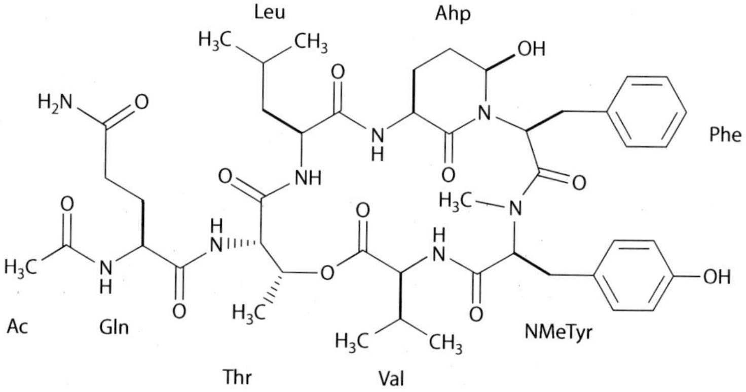

nostopeptins

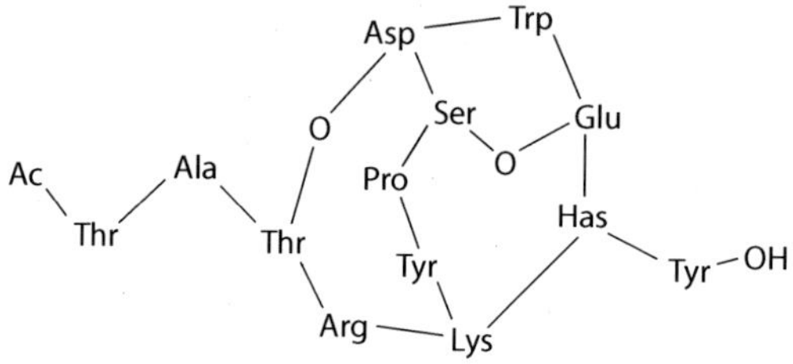

microviridin SD1634

NOSTOC SPONGIAEFORME C. AGARDH, 1824

Phylum: Cyanobacteria
Class: Cyanophyceae
Order: Nostocales
Family: Nostocaceae
Distribution: Europe, Central Asia
Habitat: Freshwater habitats, rice fields
Description: The colonies of this species are initially spherical, and later they expand and develop into humps. The filaments, which are blue-green or brownish, are fluxuous and loosely entangled.
Compounds and activities: The extract of this species has shown antimicrobial properties. The values of inhibition zone diameter (mm) for the bacterial species, *S. epidermidis*, *S. aureus*, *S. haemolyticus*, *P. vulgaris*, and *S. sonnei*, were found to be 21, 22, 15, 17, and 19, respectively. The values for the fungal species, *C. krusei*, *C. albicans*, *A. niger*, *A. fumigatus*, and *C. neoformans*, were 9, 10, 8, 6, and 9 mm, respectively (Ramamurthy et al., 2012). The bioactive compound isolated from this species, viz., borophycin, is cytotoxic acting against human epidermoid carcinoma (Lovo) and colorectal adenocarcinoma (KB) cell lines.

HAPALOSIPHON WELWITSCHII WEST & C.S. WEST, 1897

Phylum: Cyanobacteria
Class: Cyanophyceae
Order: Nostocales
Family: Hapalosiphonaceae
Distribution: Australia, New Zealand, Africa, Vietnam
Habitat: Rice fields, swamps, lakes
Description: The thallus of this species is free and forms coiled filamentous clusters attached to the substrate. The filaments of this species are irregularly arcuated with uniseriate trichomes. The cells are cylindrical, and the heterocysts, which are intercalary, are also cylindrical.
Compounds and activities: The extract of this species has been found to possess antibacterial activities. The values of inhibition zone diameter (mm) for the bacterial species, *S. epidermidis*, *S. aureus*, *S. haemolyticus*, *P. vulgaris*, and *S. sonnei*, were found to be 20, 21, 17, 19, and 18, respectively. The extract of this species containing bioactive compound welwistatin has shown antifungal properties. The values of inhibition zone diameter for the fungal species, *C. krusei*, *C. albicans*, *A. niger*, *A. fumigatus*, and *C. neoformans*, were found to be 10, 9, 6, 4, and 9 mm, respectively (Ramamurthy et al., 2012). The bioactive compound hapalosin (cyclic depsipeptide) isolated from this species has shown reversing activity against multidrug resistance (MDR), and it has direct applications as a drug delivery system. Further, this compound has also shown cytotoxic activity (Singh et al., 2005).

ANABAENA LAXA (RABENHORST) A. BAUN

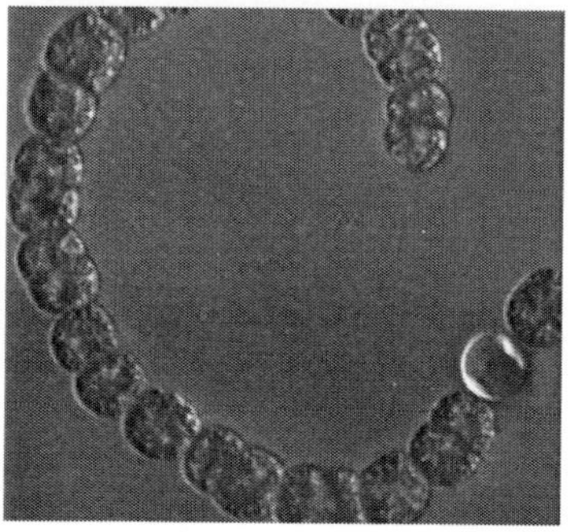

Phylum: Cyanobacteria Class: Cyanophyceae
Order: Nostocales Family: Nostocaceae
Distribution: Europe, Australia, New Zealand
Habitat: Rice fields, terrestrial habitats
Description: The cells of this species form unbranched filaments that are randomly twisted or coiled. The vegetative cells (12 × 14 mm) are spherical to oblong and are with granules. The akinetes (50–15 mm) are larger than vegetative cells. The heterocysts (10 × 9 mm) are spherical and empty.
Compounds and activities: The compounds isolated from this species are laxophycins A–E, which possess antifungal properties (Chaudhary et al., 2012).

7 A17

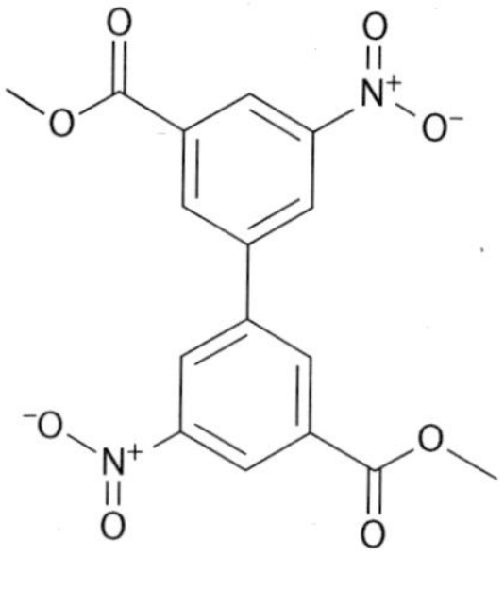

laxaphycin

ANABAENA CYLINDRICA LEMMERMANN, 1896

Phylum: Cyanobacteria Class: Cyanophyceae
Order: Nostocales Family: Nostocaceae
Distribution: Europe, Asia, Australia, New Zealand
Habitat: Freshwater ponds, lakes
Description: Cells (4 μm wide) are cylindrical and the trichome is without branching. Heterocysts are provided with a thin mucus sheath.
Compounds and activities: The extracts of this species have been found to possess antimicrobial activities. The values of inhibition zone diameter (mm) for the

bacterial species *S. epidermidis*, *S. aureus*, *S. haemolyticus*, *P. vulgaris*, and *S. sonnei* were found to be 19, 19, 16, 17, and 18, respectively. The values for the fungal species, *C. krusei*, *C. albicans*, *A. niger*, *A. fumigatus*, and *C. neoformans*, were 9, 9, 6, 5, and 8 mm, respectively (Ramamurthy et al., 2012).

TOLYPOTHRIX DISTORTA KUTZING, 1843

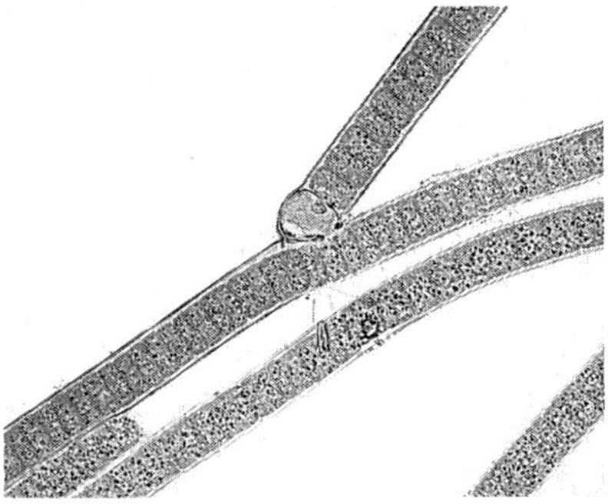

Phylum: Cyanobacteria
Class: Cyanophyceae
Order: Nostocales
Family: Microchaetaceae
Distribution: Europe, Southwest Asia
Habitat: Freshwater habitats, streams
Description: This soft-bodied species has dark blue-green or brown colonies that are mat- or cushion-like. The filaments (3 cm × 20 μm) have repeated, single false branching at the heterocyst. The cells (10 × 15 μm) are slightly constricted at cross walls. The heterocysts are in rows of two or three, but usually they are single. The sheath is initially colorless but becomes yellow-brown with age.
Compounds and activities: The bioactive compounds isolated from this species, viz., boronic acid and ethyl-dimethyl ester, have shown antibacterial activities against *Staphylococcus aureus*, *Bacillus subtilis*, and *Shigella sonnei*. A maximum activity of 83.9% was recorded with ethyl-dimethyl ester (Al-Wathnani et al., 2012).

HAPALOSIPHON FONTINALIS (C. AGARDH) BORNET, 1889

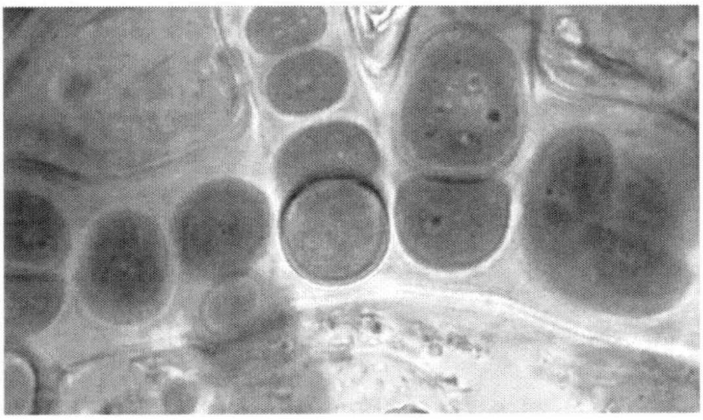

Phylum: Cyanobacteria
Class: Cyanophyceae
Order: Nostocales
Family: Hapalosiphonaceae
Distribution: Pacific Islands, Australia, New Zealand, Southwest Asia, Southeast Asia
Habitat: Freshwater habitats, paddy fields
Description: This filamentous species has several coiled filaments as clusters. The filaments are irregularly arcuated with uniserial trichomes. The sheaths are thin and colorless. The heterocysts are intercalarym and the cylindrical cells possess fine and sparse granules.
Compounds and activities: The bioactive compounds isolated from the aqueous and methanol extracts of this species include fontonamide and anhydrohepaloxindole A (alkaloids) possessing antibacterial activities and several antimycotic alkaloids possessing antifungal properties (Moore et al., 1987). The values of inhibition zone diameter (mm) for the bacterial species *S. epidermidis*, *S. aureus*, *S. haemolyticus*, *P. vulgaris*, and *S. sonnei* were found to be 18, 19, 13, 16, and 14, respectively. The values for the fungal species, *C. krusei*, *C. albicans*, *A. niger*, *A. fumigutes*, and *C. neoformans*, were 7, 9, 8, 6, 7, and 7 mm, respectively (Ramamurthy et al., 2012).

FISCHERELLA MUSICOLA (THURET) GOMONT, 1895

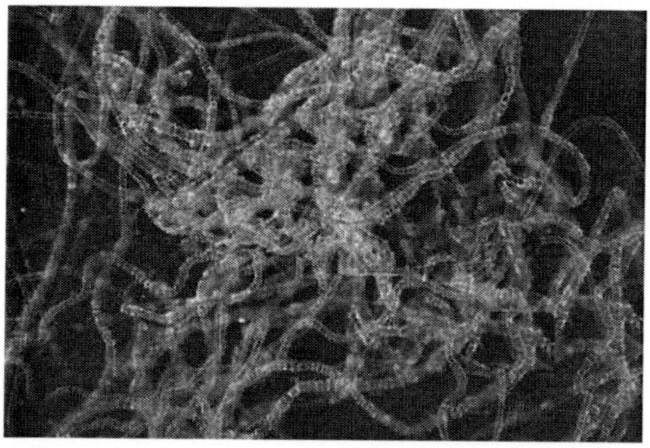

Phylum: Cyanobacteria
Class: Cyanophyceae
Order: Nostocales
Family: Hapalosiphonaceae
Distribution: Europe, Australia, New Zealand
Habitat: Freshwater habitats
Description: This benthic species has uni- and multiserial filaments. Heterocysts are present in trichomes.
Compounds and activities: The bioactive compounds fischerellin A and B isolated from this species have shown antibacterial activities against *Micrococcus luteus*. A strong inhibition of 22–-27 mm was recorded in this regard (Srivastava, 1996).

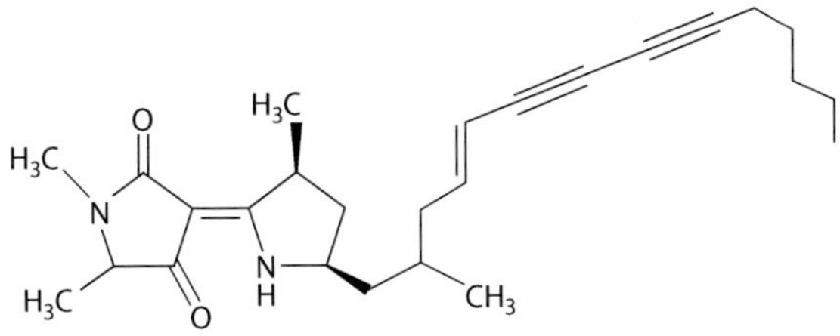

(a)

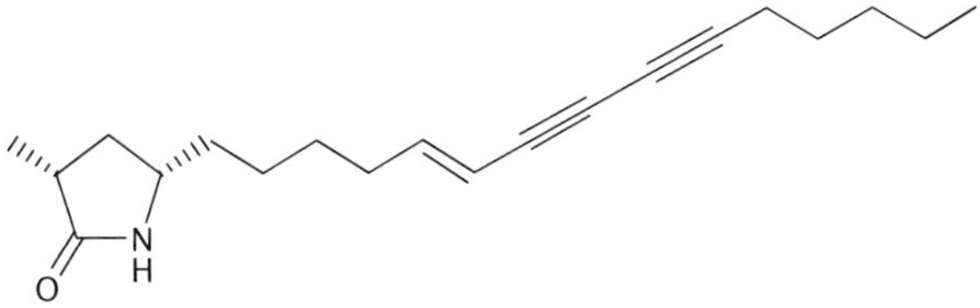

fischerellin A, B

(b)

FISCHERELLA AMBIGUA (KUTZING EX BORNET & FLAHAULT) GOMONT, 1895

Phylum: Cyanobacteria Class: Cyanophyceae
Order: Nostocales Family: Hapalosiphonaceae
Distribution: Asia, Australia, New Zealand, Pacific Islands, United States, Canada, Mexico, Iran
Habitat: Freshwater habitats, submerged substrata, paddy fields
Description: This species has uni- and multiserial filaments. The trichomes are moniliform and creeping. The cells of the filaments are with granules. The heterocysts in the basal trichomes are subspherical, but the heterocysts of the branches are cylindrical.
Compounds and activities: The bioactive compound, fisherindole L, isolated from this species has shown antifungal properties. Other compounds isolated from this species include parsiguine, fischambiguines A and B, ambiguine P, ambiguine Q nitrile, ambiguine G nitrile, and hapalindole-related alkaloids. Among them, fischambiguines A and B have shown inhibitory activities against *Mycobacterium tuberculosis*. The petroleum ether fraction of this species has, however, shown antibacterial activities particularly against *Escherichia coli*. Its ambiguine K and M isonitriles have also shown most potent activity against *M. tuberculosis*, with minimum inhibitory concentration (MIC) values of 6.6 and 7.5 μM, respectively. Ambiguine A isonitrile has shown the most potent activity against *Bacillus anthracis* with a MIC of 1.0 μM (Mo et al., 2009). The novel compound parsiguine has been reported to show antimicrobial properties (Shokravi et al., 2004).

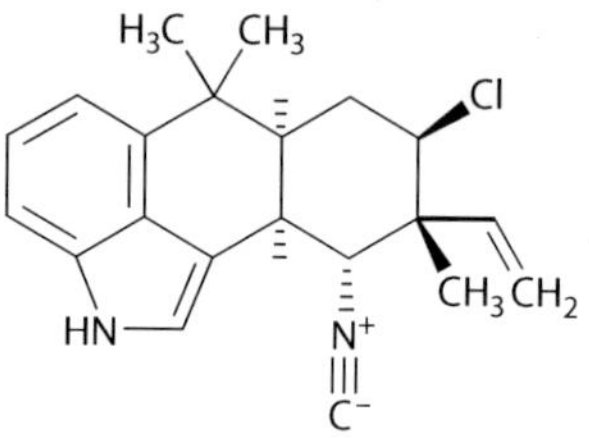

hapalindole

CHROOCOCCUS DISPERSUS (KEISSLER) LEMMERMANN, 1904

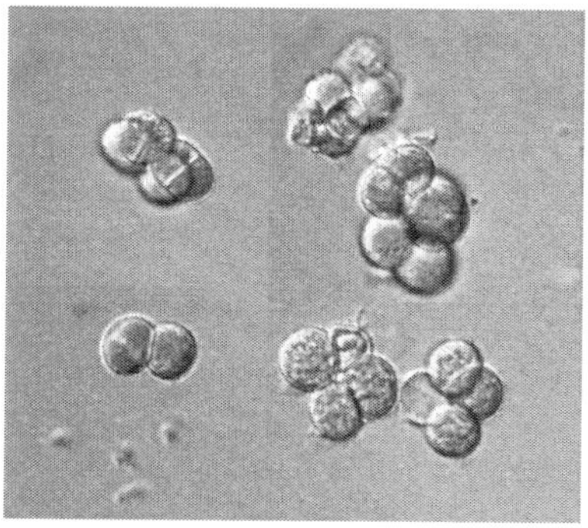

Phylum: Cyanobacteria
Class: Cyanophyceae
Order: Chroococcales
Family: Chroococcaceae
Distribution: Europe, South America, Southwest Asia, Southeast Asia, Australia, New Zealand
Habitat: Rice fields, lakes
Description: This species has small, floating colonies. Each colony may have 5–40 subspherical or elliptical cells of 3–5 μm size. The colonies may be embedded in colorless mucilage or with a narrow distinct sheath.
Compounds and activities: The methanolic extracts of this species have shown a widespread spectrum of antibacterial activities. The observed values of inhibition zone diameter (mm) for the bacterial species, *S. aureus*, *S. epidermidis*, *B. subtilis*, *E. coli*, *S. typhi*, and *P. aeruginosa*, were 10, 8, 10, 8, 10, and 8, respectively. The antifungal activity of the supernatants of this species was observed for the fungal species, *C. kefyr*, *A. fumigatus*, and *A. niger*, and the values of inhibition zone diameter were 14, 8, and 11 mm, respectively (Ghasemi et al., 2007).

CHROOCOCCUS TURGIDUS (KUTZING) NAGELI, 1849

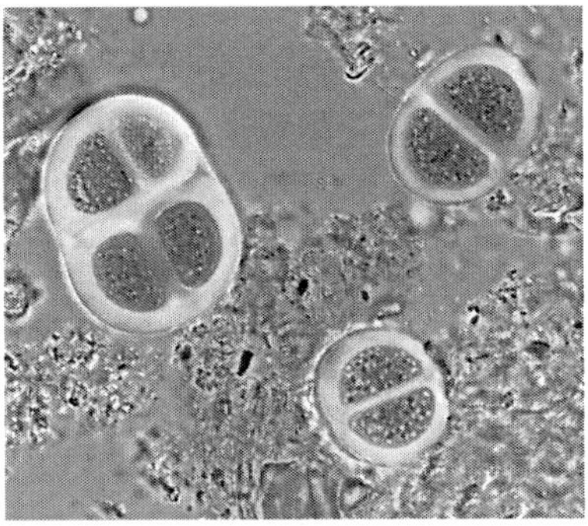

Phylum: Cyanobacteria
Class: Cyanophyceae
Order: Chroococcales
Family: Chroococcaceae
Distribution: Europe, North America, Caribbean Islands, South America, Africa, Southwest Asia, Southeast Asia, Australia, New Zealand
Habitat: Sludges of lakes and rivers
Description: This species forms microscopic colonies with 2–10 cells. The cells (10–25 µm width) are blue-green, olive-green, or pale blue-green. They are subspherical, ovoid, or hemispherical. The cells are often enveloped by colorless mucilage on colony margins.
Compounds and activities: The bioactive compounds isolated from this species include unsaturated fatty acids such as oleic and linoleic acids and unsaponiable compounds, viz., cholesterol, campesterol, stigmasterol, and β-sitosterol. These compounds have been reported to possess antibacterial activities (Abdo et al., 2012).

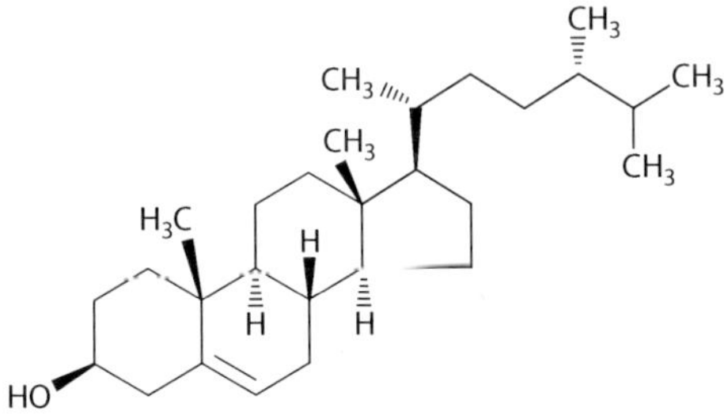

campesterol

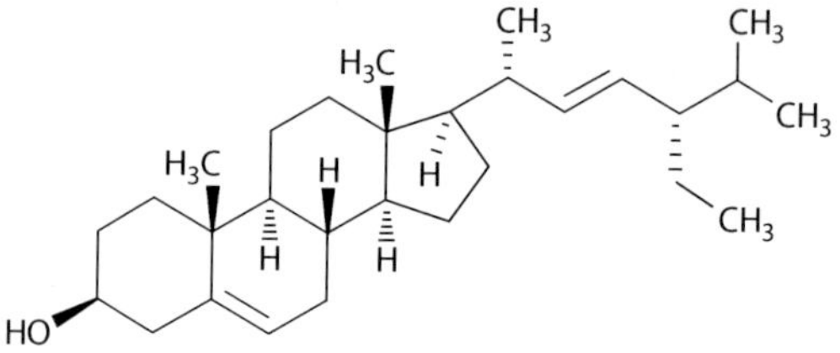

stigmasterol

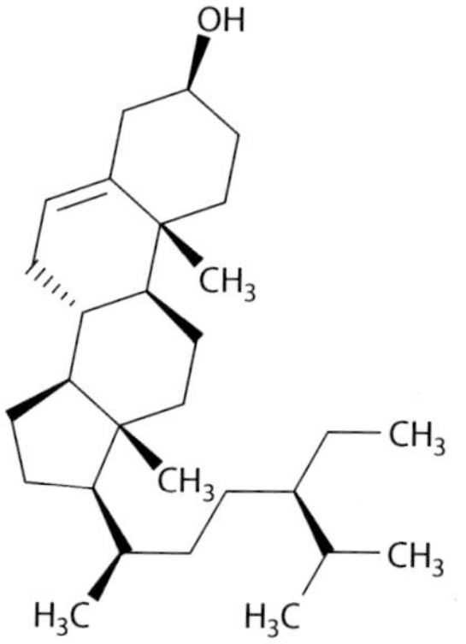

beta sitosterol

MICROCYSTIS AERUGINOSA (KUTZING) KUTZING, 1846

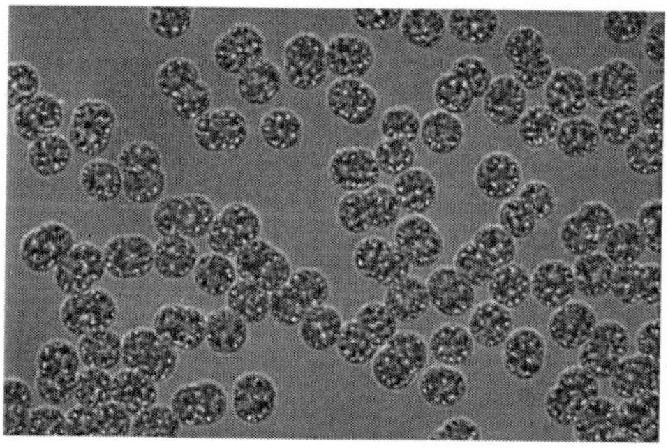

Phylum: Cyanobacteria Class: Cyanophyceae
Order: Chroococcales Family: Microcystaceae
Distribution: Central Asia, Southeast Asia, Southwest Asia, Australia, New Zealand, North America, South America
Habitat: Eutrophic freshwater habitats
Description: The small planktonic cells of this species form harmful algal blooms. These cells, which lack an individual sheath, form spherical or irregularly shaped and perforated colonies. The colonies may be light blue-green, dark, or brown due to the presence of gas-filled vesicles in their cells.

Compounds and activities: The aqueous extract of this species has been reported to possess antibacterial and antiviral (influenza A) activities. The bioactive compounds of this species include aeruginosin 98 A and B (1) (trypsin inhibitors), aeruginosin 298 A (trypsin and thrombin inhibitors), micropeptin 90 A and B (plasmin and trypsin inhibitors), micropeptin 478 A and B (plasmin, leucin, and aminopeptidase inhibitors), microginin (angiotensin-converting enzyme inhibitor), microginin 299 A and B (leucin and aminopeptidase inhibitors), microviridins B and C (elastase and serine protease inhibitors), microcystilide A (cell differentiation-promoting peptide), kawaguchipeptin B (bactericide), anabaenopeptin A ($C_{44}H_{57}N_7O_{10}$) (protease inhibitor), and (S)-(+)-1-cyclohexylethylamine (booster of mental functions). A maximum antibacterial activity of 91.9% was recorded with the last compound. (Al-Wathanani et al., 2012). A total of 12 hepatotoxins have also been isolated from this species (Namikoshi et al., 1992).

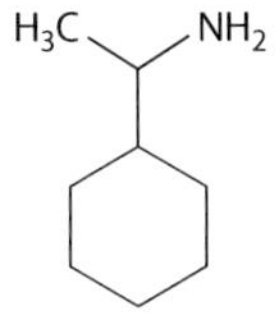

(s)-(+)-1-cyclohexylethylamine

MICROCYSTIS FLOS-AQUAE (WITTROCK) KIRCHNER, 1900

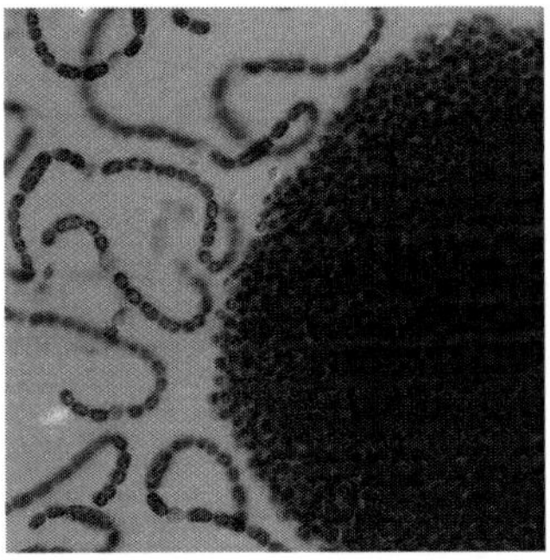

Phylum: Cyanobacteria Class: Cyanophyceae
Order: Chroococcales Family: Microcystaceae
Distribution: Europe, North America, South America, Southwest Asia, Asia, Southeast Asia, Australia, New Zealand
Habitat: Freshwater habitats
Description: This is a gas-vacuolated cyanobacterium that normally blooms in eutrophic lakes. In the free-floating colonies, the spherical cells with gas vesicles are arranged in mucilage.
Compounds and activities: A variant of known peptides isolated from the methanol/sonication extract of this species showed antibacterial activities

against *Staphylococcus aureus*, which was indicated by clear zones of 3–4 mm in bacterial lawns. Antiviral activity was also observed against dengue virus and respiratory syncytial virus (RSV) by the same extract (Victory et al., 2010).

APHANOTHECA BULLOSA (MENEGHINI) RABENHORST, 1865

Phylum: Cyanobacteria Class: Cyanophyceae
Order: Chroococcales Family: Synechococcaceae
Distribution: Europe, Southwest Asia
Habitat: Freshwater paddy fields
Description: This species forms microscopic or macroscopic colonies of several centimeters' diameter.
Compounds and activities: The methanolic crude extract (100 µg/ml) of this species was found to inhibit the growth of Gram-positive bacteria, viz., *Plesiomonas shigelloides*, *Shigella soonei*, *Bacillus* sp., *S. aureus*, *V. parahaemolyticus*, and *V. cholerae* (Kumar et al., 2012). The methanolic extracts of this species have also shown antileishmonial activity against the protozoan parasite *Leishmania donovaini*. Further, this species also possesses antifungal properties against *Candida albicans*.

OCHROMONAS DANICA E.G. PRINGSHEIM, 1955

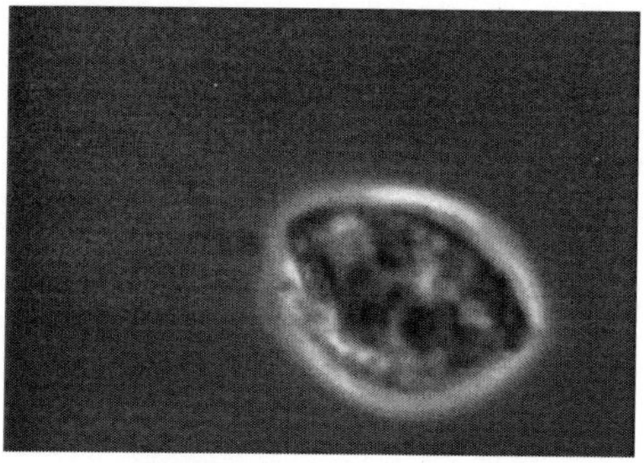

Phylum: Ochrophyta Class: Chrysophyceae
Order: Chromulinales Family: Chromulinaceae
Distribution: Cosmopolitan distribution, South America, Denmark
Habitat: Freshwater habitats
Description: This free-floating phytoflagellate is a mixotrophic bacterivore. The solitary cells (10 µm size) of this chrysophyte are naked and pyriform, spherical, or ellipsoid. This species is characterized by the presence of two unequal (long and short) flagella. The chloroplasts, numbering one or two, are yellow to brown in color.

Compounds and activities: The extracts of the freeze-dried biomass of this species have yielded 8-chlorosulfolipids, including the most common 2,2,11,13,15,16-hexachlorodocosane-1,14-disulfate (danicalipin A), which shows antibiotic properties particularly against *Flectobacillus* (Blom and Pernthaler, 2010).

POTERIOOCHROMONAS MALHAMENSIS (PRINGSHEIM) PETERFI, 1969

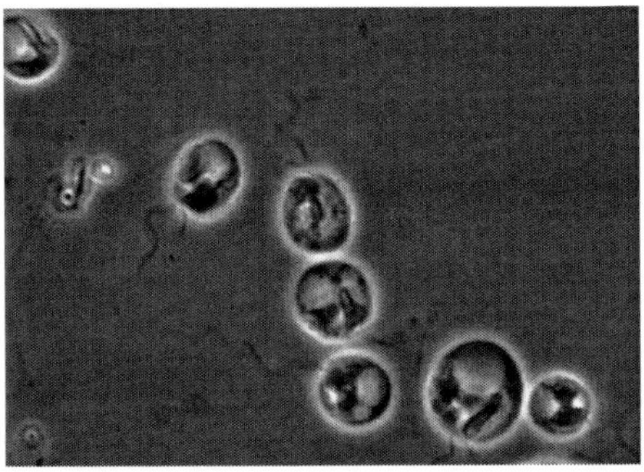

Phylum: Ochrophyta
Class: Chrysophyceae
Order: Chromulinales
Family: Dinobryaceae
Distribution: Europe
Habitat: Freshwater habitats
Description: This unicellular, mobile ochrophyte has two unequal flagella. There are vacuoplasts containing leucosin storage vacuole. The reproduction is by mitosis, cytokinesis, and the vegetative method. Lorica formation is seen in this species.
Compounds and activities: The fatty acids extracted from the freeze-dried biomass of this species have shown antibiotic properties particularly against *Flectobacillus*. Further, the bioactive compound, malhamensilipin A (disulfate-chlorolipid), of this species serves as an inhibitor of protein tyrosine kinase and shows antimicrobial and antiviral activities (Kawahara and Okino, 2012). This species also produces leucosin (= chrysolaminarin), which is a polysaccharide composed of β-(1-3)-linked d-glucose residues. Since the leucosin is a biopolymer with cellulose, it is used in the preparation of thin films for wrapping drug capsules.

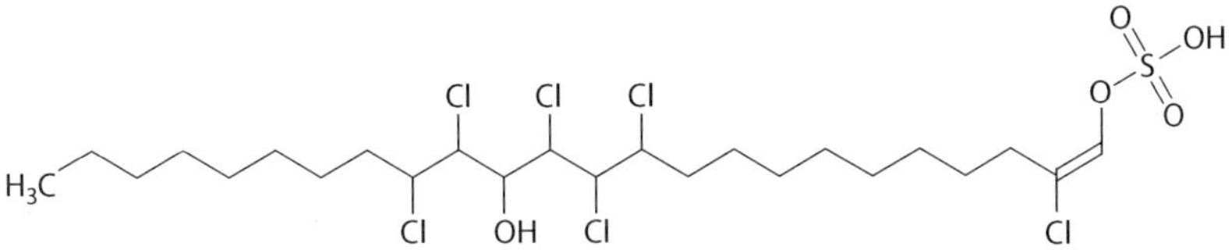

malhamensilipin A

MARSILEA QUADRIFOLIA LINNAEUS

Phylum: Pteridophyta
Class: Pteridopsida
Order: Salviniales
Family: Marsileaceae
Common name: European water fern, four-leaf clover
Distribution: Europe, Siberia, Afghanistan, India, China, Japan, North America
Habitat: Shallow-water lakes, ponds, rivers, stream edges
Description: It is an aquatic fern with four-part leaves that are floating in deep-water or erect in shallow water or on land.
Compounds and activities: The juice made from the leaves is diuretic and febrifuge. It is also used to treat snakebites and is applied to abscesses. The extracts of chloroform and ethyl acetate of these plants showed antibacterial activity against 5 Gram-positive and 11 Gram-negative human pathogenic bacterial species, and the zone of inhibition in these extracts ranged from 9 to 20 mm. These extracts also showed cytotoxicity, and the LC_{50} values were 7.820 and 8.589 μg/ml, respectively. The ethyl acetate fraction of this plant also showed potent antioxidant activity with a LC_{50} value of 50.1053 μg/ml (Ripa et al., 2009).

TETRASPORA GELATINOSA (VAUCHER) DESVAUX, 1818

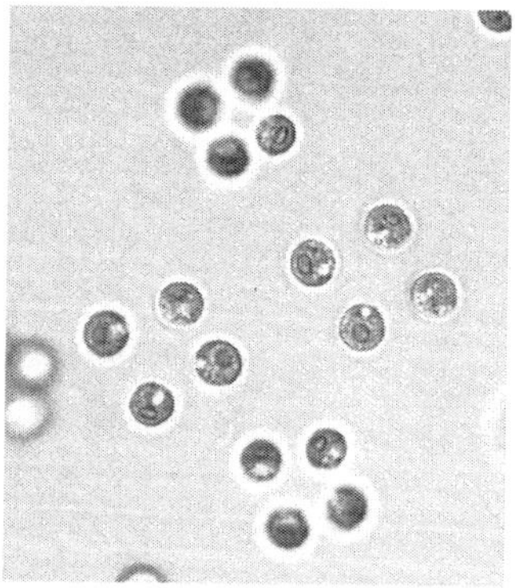

Phylum: Chlorophyta
Class: Chlorophycea
Order: Tetrasporales
Family: Tetrasporaceae
Distribution: Europe, North America, Southwest Asia, Asia, Australia, New Zealand
Habitat: Still freshwater habitats
Description: This species forms colonies that are spherical, saccate, or cumulus with a size of 1–10 cm diameter and 0.5–1 mm thickness. The cells (600 μm diameter) are green and slightly ellipsoid with a single pyrenoid.
Compounds and activities: The compounds isolated from the crude extracts of this species are pentadecatraenoic acid, 3,7,11-trimethyl-2,6,10-dodecatrienoic acid, 5,8,11-heptadecanoic acid, β-sitosterol, and transphytol. Among them, pentadecatetraenoic acid has shown antimicrobial activities against six pathogenic fungal species and three bacterial species. The percent values of activities in the first three compounds were found to be 30.8, 17.6, and 23.8%, respectively (Ghazala et al., 2004).

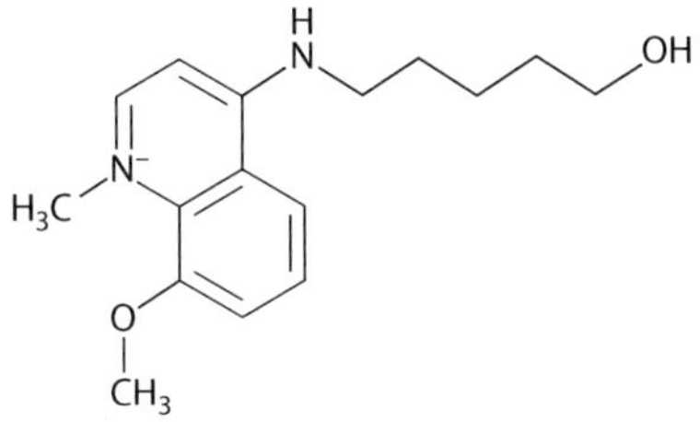

pentadecatetraenoic acid

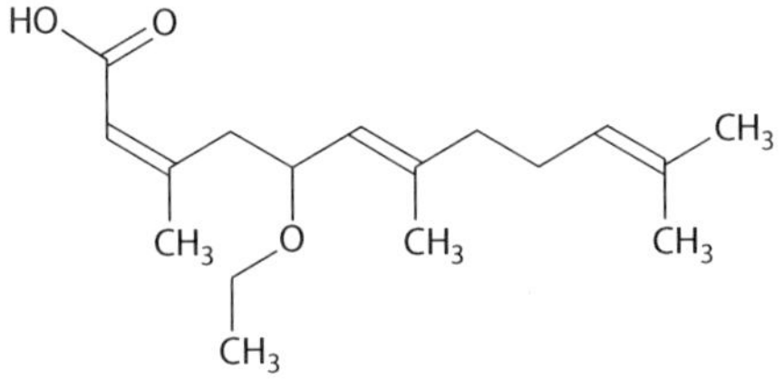

3,7,11-trimethyl-2-6,10 dodecatrienoic acid

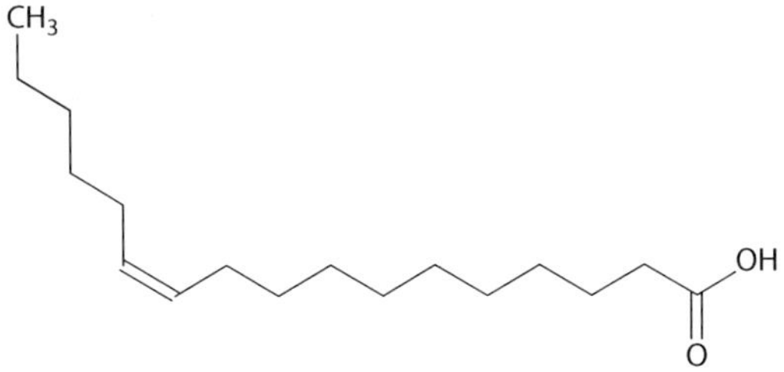

5,8,11-heptadecanoic acid

TETRASPORA CYLINDRICA (WAHLENBERG) C. AGARDH, 1824

Phylum: Chlorophyta　　Class: Chlorophycea
Order: Tetrasporales　　Family: Tetrasporaceae
Distribution: Europe, North America, Asia
Habitat: Freshwater habitats
Description: The colonies of this species may be spherical to amorphous or highly elongated, cylindrical, or irregularly shaped. The colonies may have thousands of cells, embedded in a mucilaginous matrix. The cells are spherical to oval with 5–15 μm diameter. The colonies grow by mitotic divisions and asexually by zoospores, akinetes, and colony fragmentation.
Compounds and activities: From the extracts of this species, a total of 9 saturated and 11 unsaturated fatty acids have been isolated. The important ones are β-sitosterol, transphytol, hencicosanoic acid, pentadienoic acid, and undecatrienoic acid. Among them, the last three compounds have shown antibacterial and antifungal activities. The percentage values of activities of these compounds were found to be 17.02, 0.115, and 0.115%, respectively (Ghazala et al., 2004).

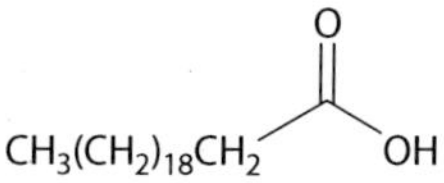

hencicosanoic acid

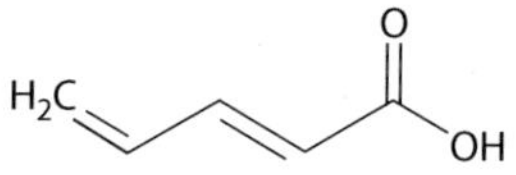

pentadienoic acid

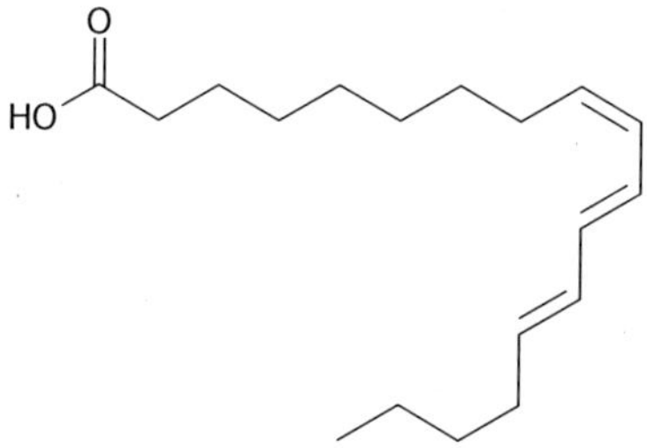

octadecatrienoic acid

CHARA GLOBULARIS J.L. THUILLER, 1799

Phylum: Charophyta
Class: Charophyceae
Order: Charales
Family: Characeae
Common name: Stone worts
Distribution: Europe, South Asia, Southwest Asia, Australia, New Zealand
Habitat: Lakes, rivers, brackish waters
Description: This monoecious, green-colored plant species, which is either annual or perennial, grows to a maximum height of only 50 cm and has 6–10 green branchlets of 7–12 segments each.
Compounds and activities: The cytokinins isolated from the extracts of this species possess antibacterial activities. The other similar bioactive compound of this species is charamin, a quaternary ammonium ion (Anthoni et al., 1987).

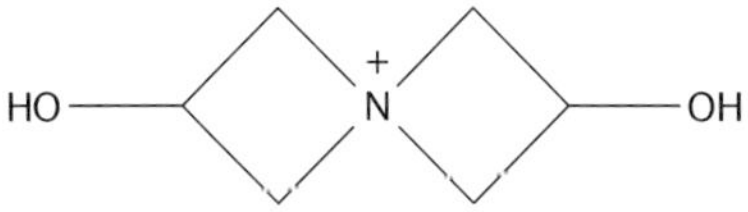

charamin

CHARA CORALLINA KLEIN EX WILLDENON

Phylum: Charophyta
Class: Charophyceae
Order: Charales
Family: Characeae
Common name: Stone wort, brittle wort, shunkweed
Distribution: Malaysia, Australia
Habitat: Ponds, pools, lakes
Description: This species, which is anchored by underground branching rhizoids, has dimorphic branches, viz., long branches of unlimited growth and short branches of limited growth. It is capable of reproducing by both sexual and vegetative methods.
Compounds and activities: From the methanolic extracts of this species bioactive compounds such as hexadecadienoic acid (23.7%), pentadecyclic acid, palmitic acid, β-sitosterol (76.8%), cholesterol, and stigmasterol have been isolated. These compounds act against *S. aureus*, *Streptococcus pyrogenes*, and *Vibrio cholerae*. The value of zone of inhibition diameter (mm) for these bacterial species was found to be 6 at a methanol extract concentration of 400 μg/ml. This extract also showed antifungal activity toward *Allescheria boydii*, *Aspertgillus niger*, *Curvularia lunata*, *Microsporum canis*, *Nigrospora oryzae*, *Pleuroetus austreatus*, *Stachybotrys atra*, and *Trichophyton mentagrophytes* (Khaliq-uz-Zaman, 1998).

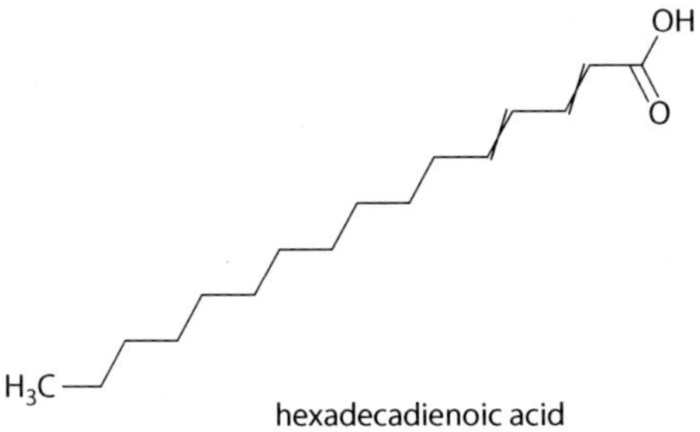

hexadecadienoic acid

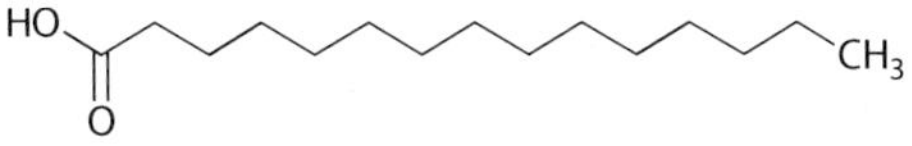

pentadecylic acid

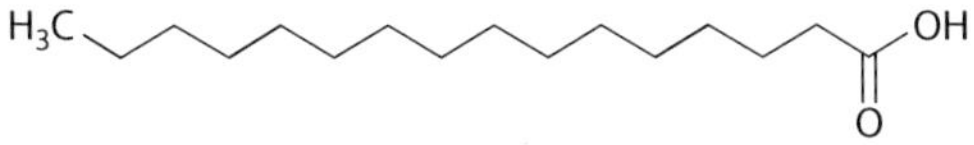

palmitic acid

CHARA ZEYLANICA KLEIN EX WILD

Phylum: Charophyta
Class: Charophyceae
Order: Charales
Family: Characeae
Common name: Stonewort
Distribution: Worldwide distribution
Habitat: Ponds, ditches, lakes, channels
Description: This monoecious species has a characteristic corticated axis, numerous branchlets, and two rows of well-developed stipulodes.
Compounds and activities: The extracts of this species contain 13 saturated and 23 unsaturated fatty acids, 2 sterols, 3 terpenes, and 1 carbohydrate.

Important bioactive compounds isolated from this species include lawsaritol, 5α-stigmast-9(11)-en3β-ol(2), 12-hydroxy-13-methylpodocarpa-9,11,13-trien-3-one(3), 12-dihydroxy-3,7,11,15-tetramethyl-1,10,14-hexadecatrien-5,13-dione, and 3,12-dihydroxy-3,7,11,15-tetramethyl-1,6,10,1-hexadeca-tetraen-5,13-dione. All these compounds have been reported to possess antibacterial activities against *Pseudomonas aeruginosa*. The alcohol, acetone, and chloroform extracts of this species showed values of zone of inhibition (mm) of 12, 11, and 8.5, respectively (Patil et al., 2010).

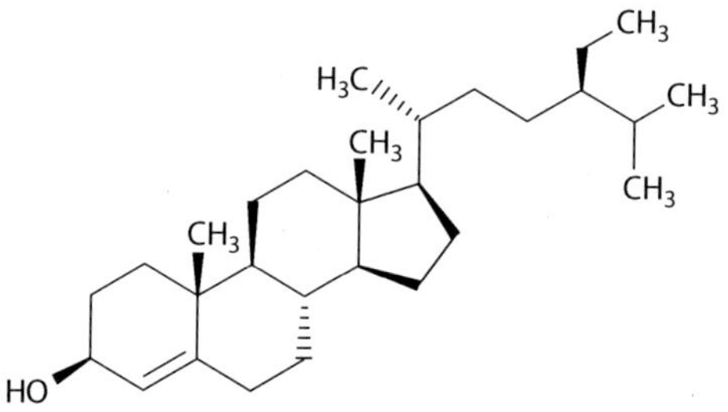

lawsaritol

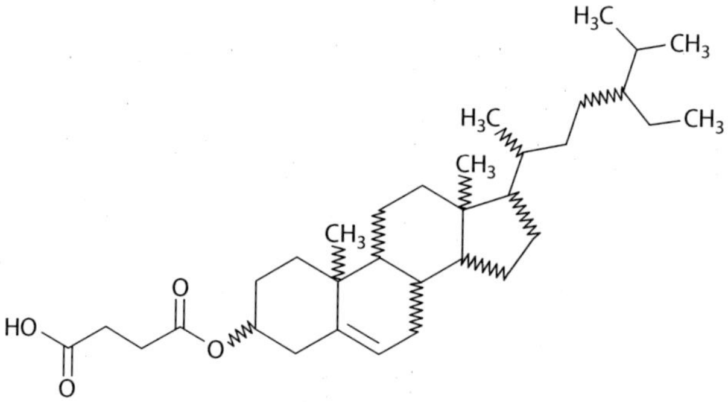

5 alpha-stigmast-9(11)-en 3beta-ol

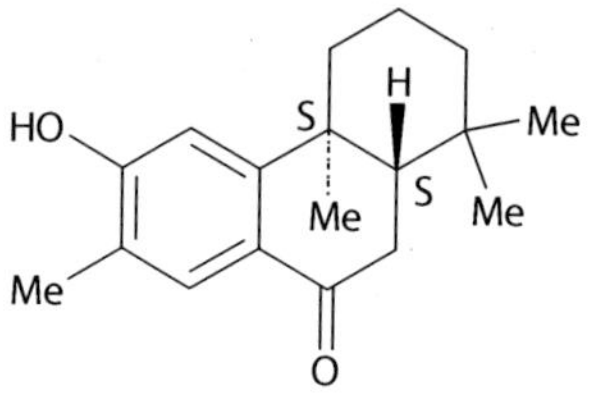

12-hydroxy-13-methyl podocarpa-9,11,13-trien-3-one

NITELLA HYALINA (DE CANDOLLE) C. AGARDH, 1824

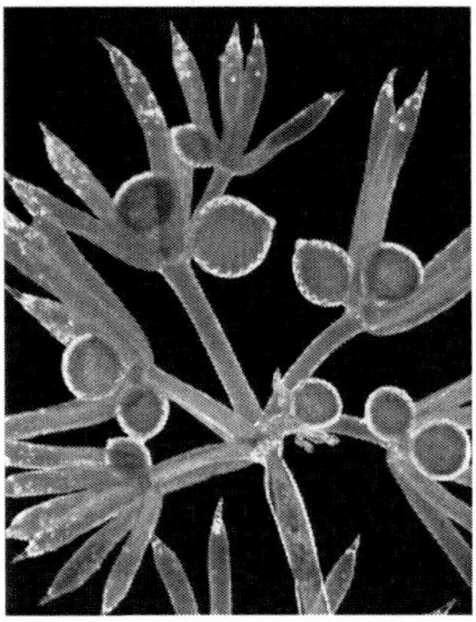

Phylum: Charophyta Class: Charophyceae
Order: Charales Family: Characeae
Common name: Many-branched stone wort
Distribution: Cosmopolitan distribution, Europe, South America, Southwest Asia, Central Asia, Pacific Islands, Australia, New Zealand
Habitat: Freshwater, brackish water, marine habitats
Description: This annual species, which grows to a maximum height of only 20 cm, has an axis measuring 800 μm in diameter. Accessory branchlets are present and the flowers are monoecious.
Compounds and activities: The extracts of this plant species have yielded 13 saturated and 23 unsaturated fatty acids, 4 sterols, 1 terpene, and 2 carbohydrates. Important bioactive compounds isolated from this species include cholesterol, ergosterol, stigmasterol, β-sitosterol, transphytol, xylomacin, and stigmasteryl galactoside. All these compounds show antimicrobial activities against seven human pathogenic species (Khalid et al., 2010).

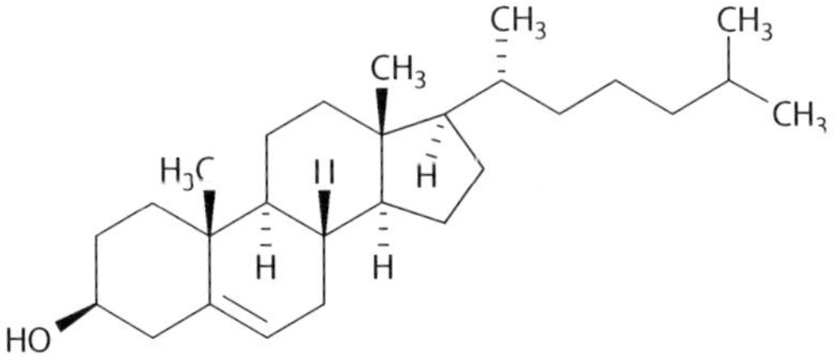

cholesterol

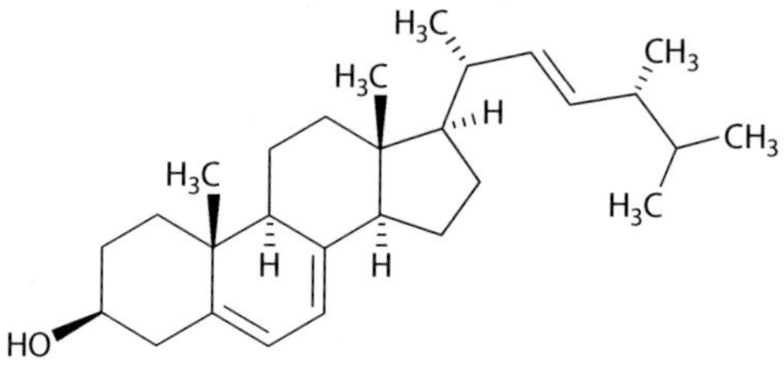

ergosterol

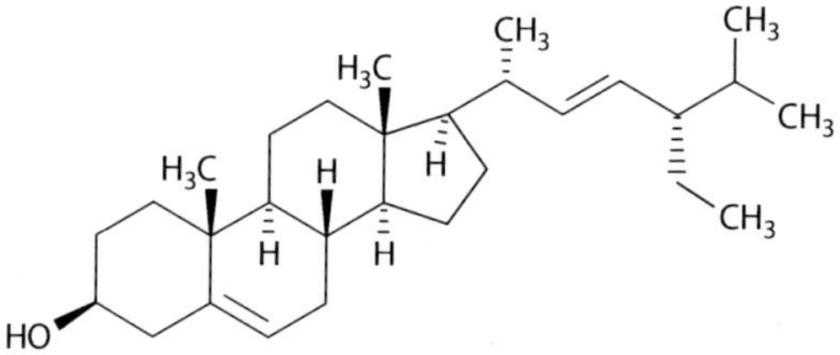

stigmasterol

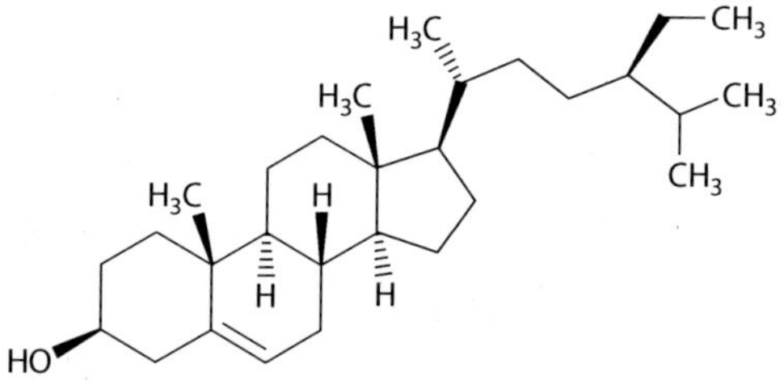

beta sitosterol

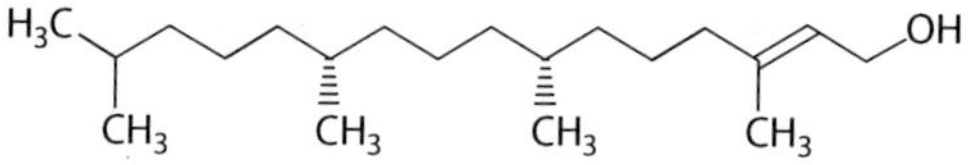

trans-phytol

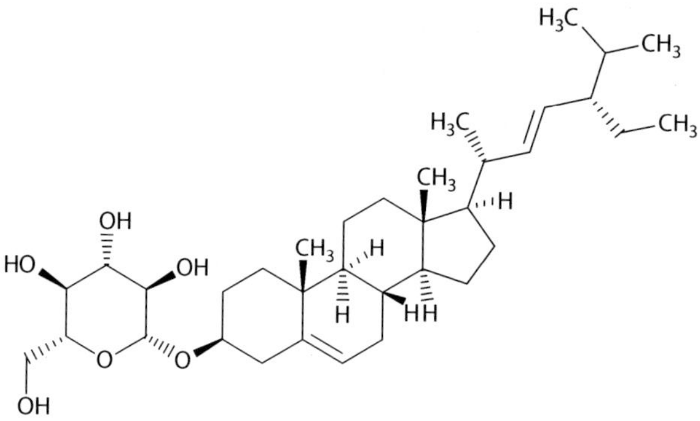

stigmasteryl galactoside

NITELLA FLEXILIS (LINNAEUS) C.A. AGARDH, 1824

Phylum: Charophyta
Class: Charophyceae
Order: Charales
Family: Characeae
Common name: Smooth stonewort
Distribution: Europe, North America, South America, Asia, Australia, New Zealand
Habitat: Lakes, ponds, ditches, rivers
Description: It is a robust monoecious species growing to a maximum height of 1 m. Branches are in whorls, which divide once or twice.
Compounds and activities: From this species come five saturated, seven mono-, two di-, two tri-, and three polyunsaturated fatty acids (FAs). The methanol extracts of this species showed antibacterial activity toward *Shigella boydii* (8 mm), *Bacillus cereus* (6 mm), and *Corynebacterium diphtheriae* (6.5 mm), and antifungal activity toward *Curvularia lunata* (3%), *Fusarium sporotrichoids* (3.5%), *Macrophomina phaseolina* (3.7%), *Rhizoctonia solani* (3.6%), *Sclerotium rolfsii* (3.4%), and *Trichoderma harzianum* (3.0%) (Ghazala et al., 2004).

EUGLENA VIRIDIS (O.F. MULLER) EHRENBERG

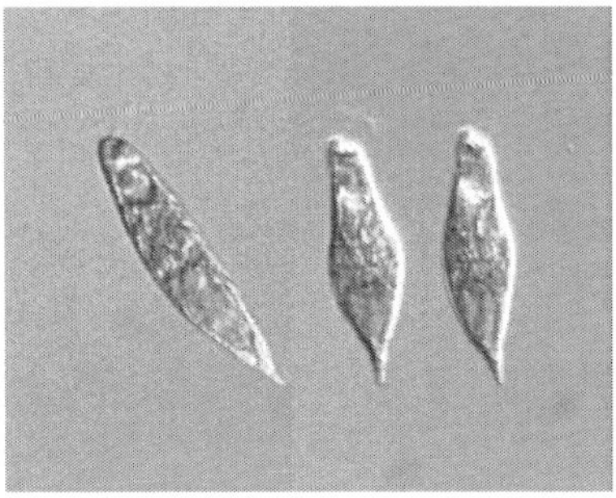

Phylum: Euglenoidea
Class: Mastigophora
Order: Euglenoidida
Family: Euglenoididae

Distribution: Australia, New Zealand, New South Wales

Habitat: Ponds, ditches, stagnant pools

Description: These green unicellular flagellates (protists or green algal protozoans) are photosynthetic with radially arranged chloroplasts and heterotrophic like animals. The biflagellated cells (80 × 20 μm) of this species, which are rounded anteriorly and pointed posteriorly, are fusiform during locomotion and are highly plastic when stationary. The nucleus of the cell is posteriorly placed.

Compounds and activities: The organic extracts of this species have shown antibacterial properties against virulent pathogens like *Pseudomonas putida*, *P. aeruginosa*, *P. fluorescens*, *Aeromonas hydrophila*, *Edwardsiella taida*, *Vibrio anguillarum*, *V. fluviatilis*, *V. parahaemolyticus*, *V. harveyi*, and *Escherichia coli* (Amaro et al., 2011).

XYRIS INDICA LINNAEUS

Phylum: Magnoliophyta Class: Magnoliopsida

Order: Poales Family: Xyridaceae

Common name: Indian yellow-eyed grass, tall yellow-eyed grass

Distribution: Indian continent, South China, Southeast Asia, Malaysia to Australia

Habitat: Swamps, streams, pools, ponds, rice fields

Description: This perennial herb is partially submerged. The stem is erect and has a maximum height of 150 cm. Roots are white or brown and fibrous. Leaves (2 × 2 cm), which are in a rosette, are sessile, narrow, alternate, and entire. Flowers (2 cm across) are hermaphroditic, yellow, sessile, and clustered and are grouped together in a terminal inflorescence.

Compounds and activities: The bioactive compounds isolated from this species are xyridin A and B. Xyridin A was saponified to the corresponding ketoacid, which on reduction furnished the (±)-3,4-dihydro-6,7-methylenedioxy-3-propylisocoumarin. All these compounds were found to possess antibacterial and antifungal properties (Saeed, 2003).

ALTERNANTHERA SESSILIS (L) R. BR. EX DC.

Phylum: Magnoliophyta Class: Magnoliopsida
Order: Caryophyllales Family: Amaranthaceae
Common name: Sessile joy weed, dwarf copper leaf
Distribution: World distribution
Habitat: Ditches, damp places, wet headlands
Description: This perennial herb, which grows to a height of 50 cm, has prostrate stems. Leaves (15 × 3 cm) are elliptic or obovate. The flowers are in sessile spikes.
Compounds and activities: The chloroform extracts of air-dried leaves of this species possess a mixture of diastereomers of a new Ionone derivative, which show antimicrobial activities against *Pseudomonas aeruginosa* and *Trichophyton mentagrophytes* and large amounts of carotenes. The crude extracts of this species have also been reported to possess cytotoxic (against three cancer cell lines), antioxidant, hematinic (hemoglobin and serum ferritin enhancement), and antilipidemic (anticholesterolemic) activities (Jalalpure et al., 2008; Arallado and Osi, 2010). In traditional medicines, this plant is used for treating gastrointestinal problems. The methanol extract of the leaves of this species possesses nootropic potential (memory-enhancing effect).

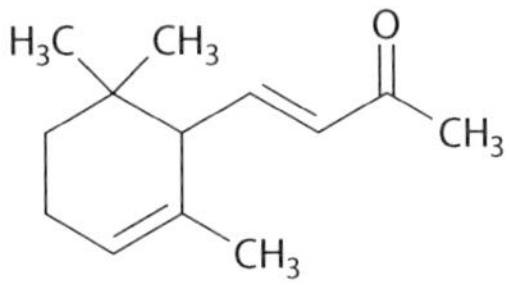

ionone

RUMEX VERTICILLATUS LINNAEUS

Phylum: Magnoliophyta
Class: Magnoliopsida
Order: Caryophyllales
Family: Polygonaceae
Common name: Swamp dock
Distribution: United States, Mexico, Canada, Europe
Habitat: Edges of pools and streams, swamps, marshes, roadside ditches
Description: This perennial unbranched plant, which grows to 1.5 m, has a central green to reddish green stem. The leaves (30 × 5 cm) are alternately arranged. The inflorescence is of one to four whorled racemes of greenish flowers, and each raceme is 30 cm long.
Compounds and activities: The bioactive compounds isolated from this species include 1-O-β-D-glucopyranosyl chrysophanal and 1-O-β-D-glucopyranosyl emodin (Kang et al., 1997). The methanol extracts of seeds of this species possess antimicrobial activities against *Staphylococcus aureus*, *Pseudomonas aeruginosa*, *Escherichia coli*, and *Candida albicans*. The phenolic compounds and tannins isolated from this species have, however, shown antioxidant properties. The antioxidant activity of the seeds of this species has been tested using trolox extract (TE), and the value recorded was 123,423 μm TE (Borchardt et al., 2009).

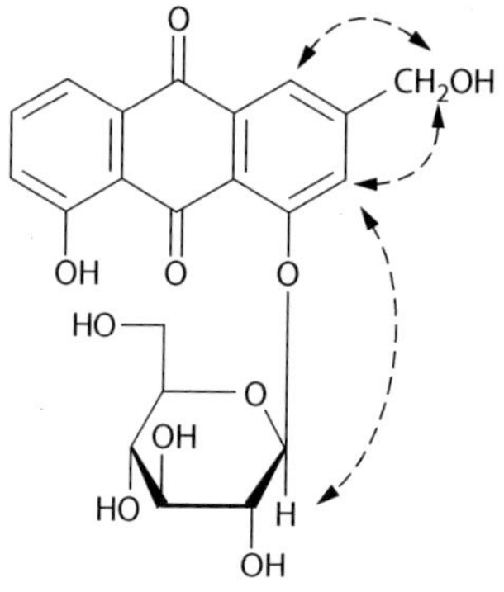

1-O-β-D-glucopyranosyl chrysophanal

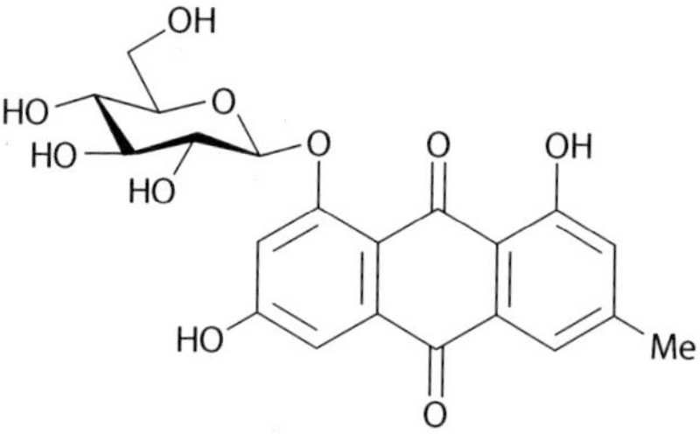

1-O-β-D-glucopyranosyl emodin

TRAPA NATANS LINNAEUS

Phylum: Magnoliophyta Class: Magnoliopsida
Order: Myrtales Family: Trapaceae
Common name: Water chestnut, water caltrop
Distribution: Europe, Asia, Africa
Habitat: Freshwater habitats (5 m deep)
Description: This is a floating annual plant species, which is anchored by very fine roots. The submersed stem is 4 m long and has two types of leaves, viz., submerged and floating. While the submersed leaves are divided and feather-like, the floating undivided leaves (2–3 cm long) are ovoid or triangular and are borne in a rosette at the water surface. The white flowers are four petaled.
Compounds and activities: The bioactive compounds isolated from this species include phenolic, flavonoid, and tannin compounds. 1,4-Dioxan extract of fruit rind of this species showed maximum antimicrobial activity. The values of inhibition zone (mm) for the Gram-positive bacterial species, *Bacillus cereus*, *B. megaterium*, *Corynebacterium rubrum*, *Micrococcus flavus*, *Staphylococcus aureus*, *S. epidermididis*, and *S. subflava*, were found to be 13.5, 10, 12, 20, 13, 12, and 11, respectively. The values with respect to Gram-negative bacterial species, *Alcaligenes faecalis*, *Klebsiella aerogenes*, *K. pneumonia*, *Proteus mirabilis*, *P. morganii*, *Pseudomonas putida*, and *P. testosteroni*, were 13, 13, 21, 12, 18, 19, and 22 mm, respectively. The values for *Aspergillus candidus*, *A. flavus*, and *Mucor hiemalis* were 15, 12, and 10 mm, respectively (Parek and Chanda, 2007). The aqueous extracts of fruits of this species have also shown antioxidative properties against free radicals, and the methanolic root extracts of this species have shown analgesic properties.

LYTHRUM ALATUM PURSH

Phylum: Magnoliophyta
Class: Magnoliopsida
Order: Myrtales
Family: Lythraceae
Common name: Winged loose strife
Distribution: United States
Habitat: Wet meadows, stream, pond margins
Description: This is a perennial shrub or subshrub, and its erect stem is 1 m tall with glabrous leaves (5 × 1 cm), which are oblong and opposite to subopposite. The inflorescence is single or paired flowers (5 × 2 m) that are pinkish purple or rose-pink with six petals.
Compounds and activities: The methanol extracts of seeds of this species have shown antimicrobial activities against *Staphylococcus aureus*, *Pseudomonas aeruginosa*, *Escherichia coli*, and *Candida albicans*. The antioxidant activity of the seeds of this species has been tested using trolox extract (TE), and the value recorded was 206,154 µm TE (Borchardt et al., 2009).

LUDWIGIA DECURRENS WALTER AND *LUDWIGIA ABYSSINICA* (A. RICH) DANDY & BRENAN

Phylum: Magnoliophyta
Class: Magnoliopsida
Order: Myrtales
Family: Onagraceae

Ludwigia decurrens

Common name: Winged water primrose, wingleaf primrose-willow

Distribution: Central and Eastern United States, Africa, Europe, temperate and tropical Asia

Habitat: Wet habitats

Description: It is an erect and perennial herb growing to a maximum height of 2 m. The linear leaves are alternately arranged and the flowers are with four or five petals each.

Ludwigia abyssinica

Common name: Water primrose

Distribution: Tropical Africa, South Africa, Madagascar

Habitat: Swampy places

Description: It is a perennial herb growing to a maximum height of 3 m. Its leaves are elliptic or lanceolate and the yellow flowers are in clusters.

Compounds and activities: The n-butanol extracts of leaves of these species have alkaloids and tannins as bioactive compounds, which possess antibacterial properties against *Escherichia coli* and *Staphylococcus aureus* with a maximum zone of inhibition (32 mm). The minimum inhibitory concentrations (MICs) exhibited by this extract against the said bacterial species range between 0.625 and 5.0 mg/ml. The killing rates of minimum bacterial concentration (MBC) of n-butanol extract of these species on *E. coli* and *S. aureus* were 99.3 and 98.2%, respectively, in 120 min (Oyedeji et al., 2011). These species are also used in traditional medicines for the treatment of skin, gastrointestinal, and bone joint disorders. While the leaf sap is used for pregnancy as an antiabortifacient, the root is used as a vermifuge.

LUDWIGIA ALTERNIFOLIA LINNAEUS

Phylum: Magnoliophyta Class: Magnoliopsida
Order: Myrtales Family: Onagraceae
Common name: Bushy seed box, Rattle box
Distribution: North America
Habitat: Wet ground, borders of streams and bodies of water
Description: This erect and branching herbaceous perennial plant grows to a height of 1 m. The alternate leaves are lanceolate and sessile or short petiolate. Inflorescence is single axillary flowers that are yellow to white with four petals each.
Compounds and activities: The bioactive compounds isolated from this species include shikimic acid, ellagitanin, strictinin, and C-glycosidic flavonoids, viz., vitexin, isovitexin, orientin, and isoorientin, which possess antimicrobial properties (Rao and Merugu, 2012).

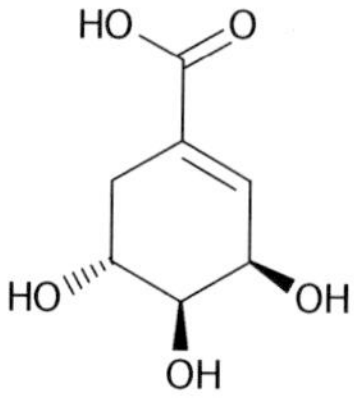

shikimic acid

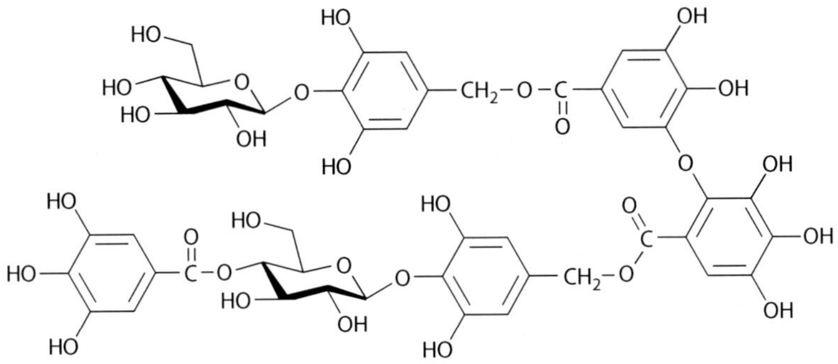

ellagitannin

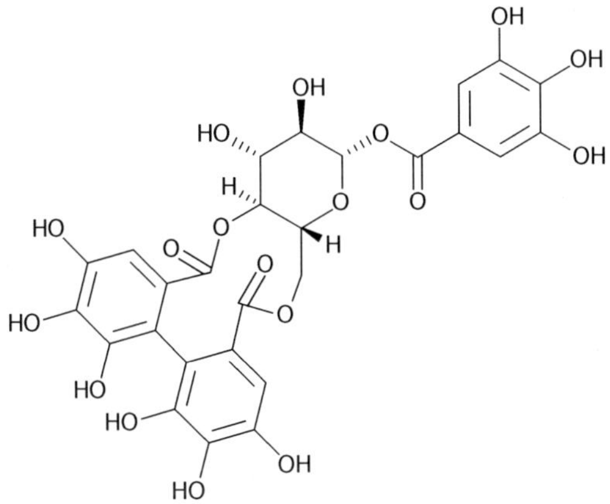

strictinin

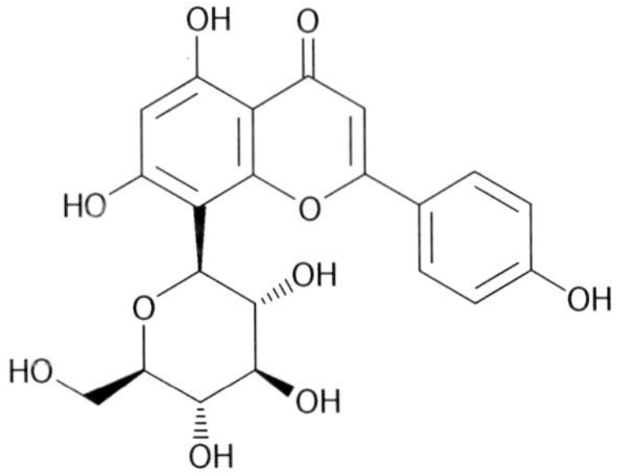

vitexin

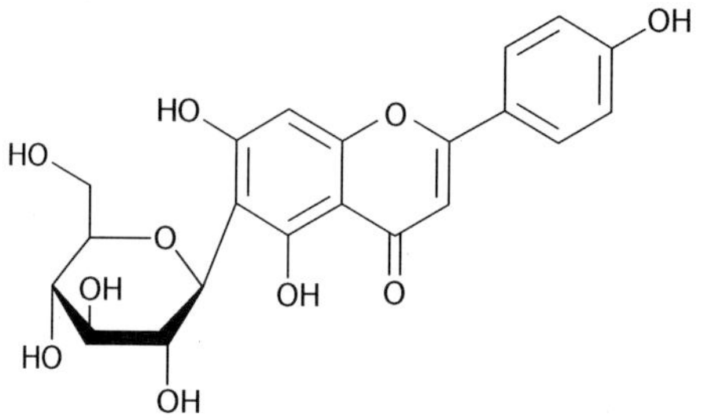

isovitexin

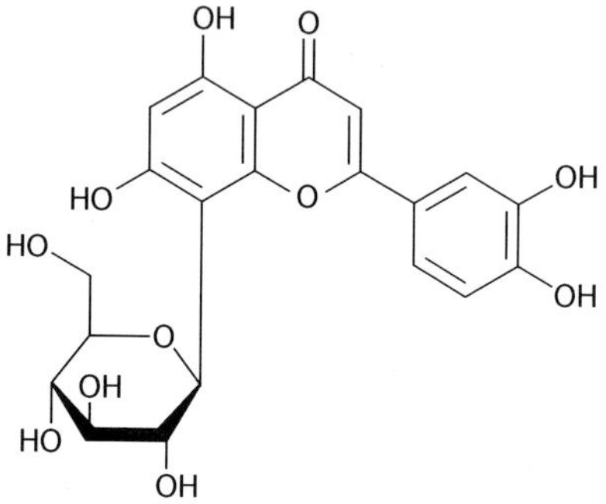

orientin

LUDWIGIA OCTOVALIS (JACQ)

Phylum: Magnoliophyta
Class: Magnoliopsida
Order: Myrtales
Family: Onagraceae
Common name: Primrose willow, swamp primrose, water primrose
Distribution: Hawaii and Pacific Islands
Habitat: Wet places, marshes, lakes, streams
Description: This is a well-branched, stout, and erect perennial or annual herb that may grow at times as a shrub of 4 m in height. The leaves are lanceolate to oval and alternately arranged. Flowers are solitary in the axils of leaves.
Compounds and activities: The oleonone-type triterpenes isolated from this species include (23Z)-feruloyl hederagenin, (23E)-feruloyl hederagenin, β-amyrin acetate, and β-amyrin palmitate (Chang and Kuo, 2007). The extracts of this plant species possess antioxidant, antidiabetic, and immunosuppressive properties. The methanol leaf (80%) extract gave the lowest minimum inhibitory concentration (62.5 μg/ml) and minimum bactericidal concentration (125.0 μg/ml) against *E. coli* 0157:H7, *E. coli* ATCC 25922, and *Bacillus spizizenii* (ATCC 6633) (Yakob et al., 2012). This species has also shown antibacterial activities against *Escherichia coli*. Further,

a highest antioxidant activity (evaluated by 2,2-diphenyl-1-picrylhydrazyl and ferric-reducing antioxidant power assays) was observed in 80% methanol extract of the leaf. The values in these assays were found to be 1,080.84 ± 6.07 μm TE/mg dry weight and 1,256.88 ± 5.38 μm TE/mg dry weight, respectively. Its chlorophyll *a* constant has shown antiproliferative properties. In traditional medicines, this plant is used for the treatment of skin disease, diarrhea, and flatulence.

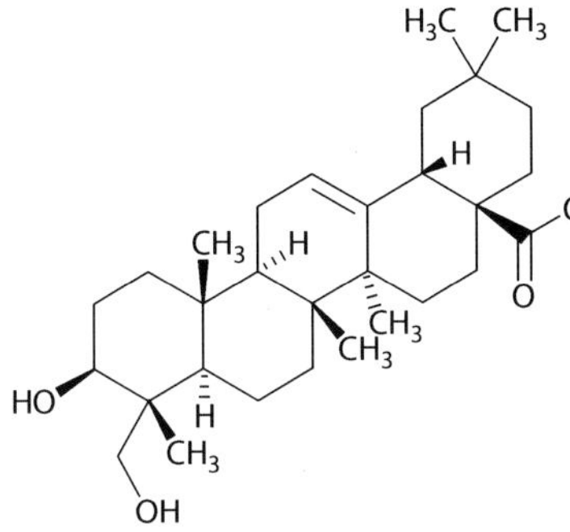

(23Z) feruloyl hederagenin

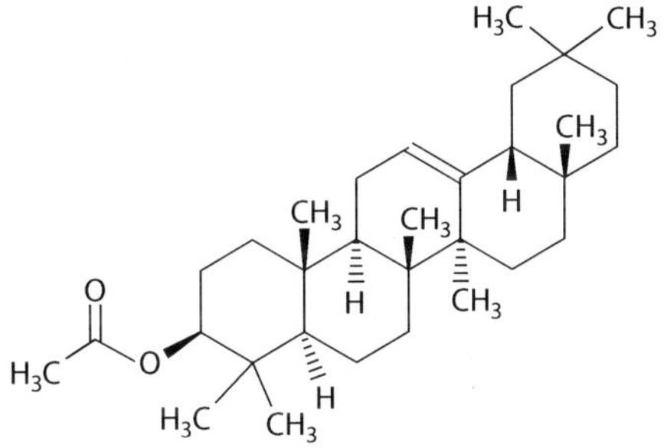

β-amyrin acetate

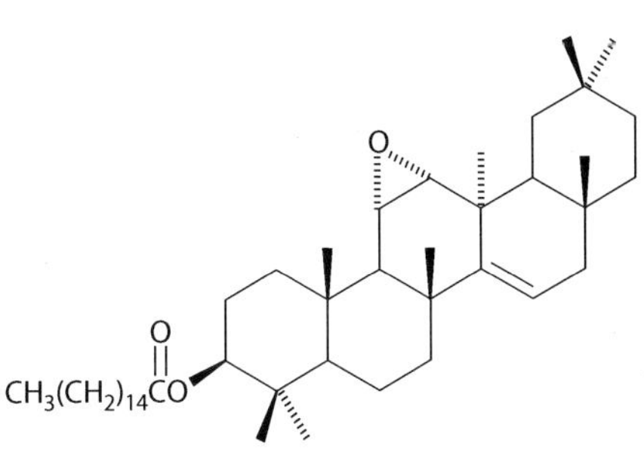

β-amyrin palmitate

LUDWIGIA ADSCENDENS (LINNAEUS) HARA

Phylum: Magnoliophyta
Class: Magnoliopsida
Order: Myrtales
Family: Onagraceae
Common name: Water primrose
Distribution: Bangladesh
Habitat: Ponds, ditches

Description: This floating herb has rooting and clusters of white, spindle-shaped pneumatophores at the nodes. The leaves, which are obovate or oblanceolate and obtuse, are 2–8 cm long. The flowers are white with ovate petals 1.5 cm long.

Compounds and activities: The stem and leaf extracts of this species have several bioactive compounds with antimicrobial and anti-inflammatory activities. The compounds isolated from this species include flavonoid glycosides, viz., kaempferol-3-glucoside, quercetin-3-O-rhamnoglucoside (rutin), quercetin-3-galactoside, mericetin-3-rhamnoside, and isosalipurposide. Other compounds isolated include terpenes, triterpenoids, flavonoids, anthroquinones, phenols, tannins, alkaloids, and ursolic acid. The microbial inhibitory activities of this plant relate to *Proteus vulgaris* and *S. aureus*. The MIC values of stem aqueous extract and leaf methanol extract of this species against *Proteus vulgaris* were found to be 2 and 0.5 mg/ml, respectively. On the other hand, the MIC values of stem hexane extract, stem methanol extract, leaf ethanol aqueous extract, and leaf methanol extract of this species against *Staphylococcus aureus* were found to be 0.5, 0.031, 1, and 0.031 mg/ml, respectively (Jani et al., 2012).

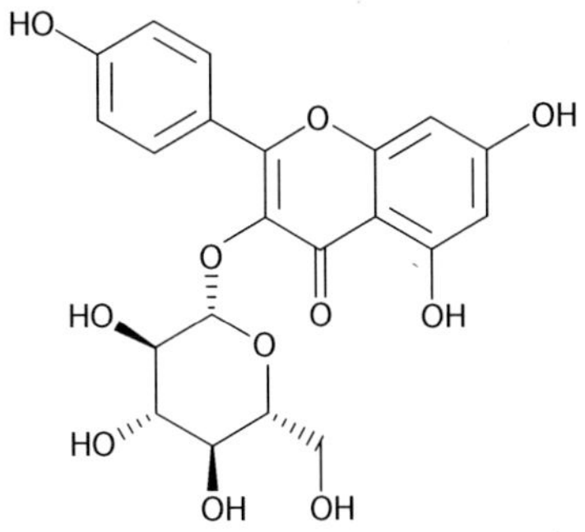

kaempferol-3-glucoside

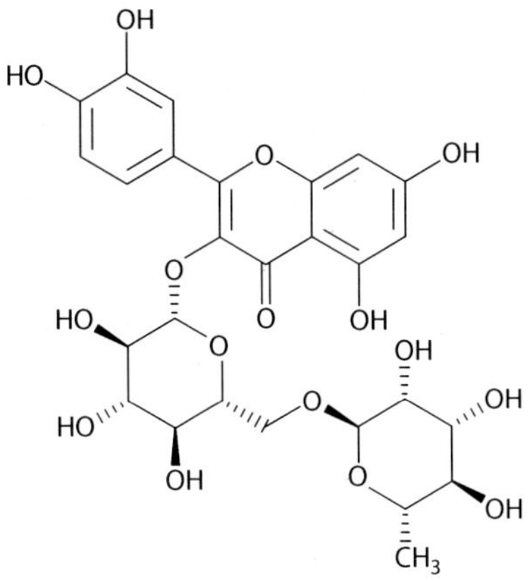

quercetin-3-O-rhamnoglucoside (rutin)

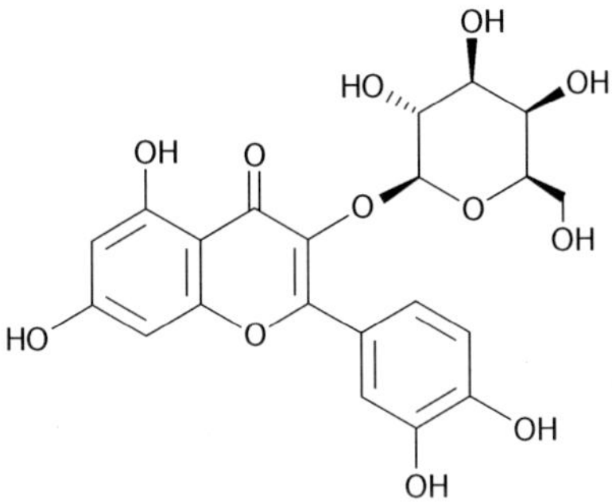

quercetin-3-galactoside

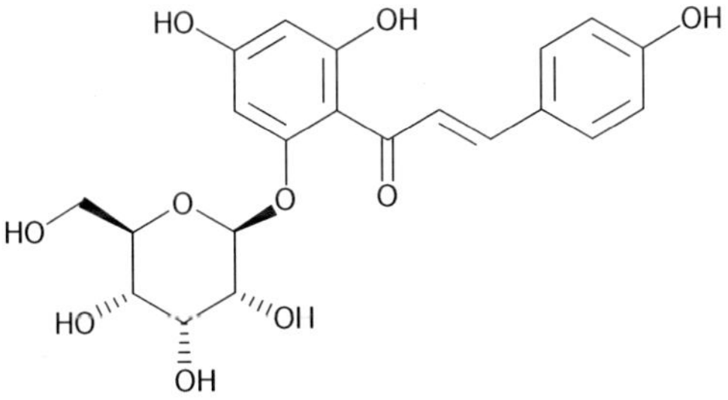

isosalipurposide

SIDA CORDIFOLIA LINNAEUS

Phylum: Magnoliophyta
Class: Magnoliopsida
Order: Malvales
Family: Malvaceae
Common name: Bala, flannel weed
Distribution: Tropical regions of the world, Africa, Australia, United States, India, Nepal
Habitat: Freshwater, wetlands, moist places, spoilt soils
Description: This is a perennial shrub growing to a maximum height of 1 m. The stem is erect and its leaves are cordate, hairy, and with serrated edges. The yellow flowers are found gathered in final inflorescences.
Compounds and activities: The methanol extract of this species was found to be effective against all phytopathogens with low MIC values of 6 µg/mm, and it showed higher inhibition activities against different bacterial species (Joseph et al., 2011). The root and stem extracts of this plant species possess bioactive compounds, such as vascinol, ephedrine, vascicone, and hypaphorine, having antimicrobial activities. They also show cytotoxic effects on HeLa cell lines. Other compounds isolated from the aerial parts of this species include pseudoephedrine and ephedrine, traces of sitosterol, palmitic, stearic, and hexacosanoic acids, 5,7-dihydroxy-3-isoprenyl flavone, 5-hydroxy-3-iso-prenyl flavones, β-sitosterol and stigmasterol (flavones), and 5′-hydroxymethol-1′-(1,2,3,9-teytrahydropyrrolo(2,1-b)quinazolin-1-yl)-heptan-1-one (analgesic alkaloid).

The seed oil of this species has sterculic, malvaic, and coronaic acids along with other fatty acids. The 50% ethanol extracts of this plant have also been reported to possess anti-inflammatory, antioxidant, antiproliferative, antipyretic, antinociceptive, antiulcerogenic, and hepatoprotective activities. In folk medicines, this plant is used for the treatment of inflammation of oral mucosa, asthmatic bronchitis, blenorrhea, nasal congestion, and other respiratory problems.

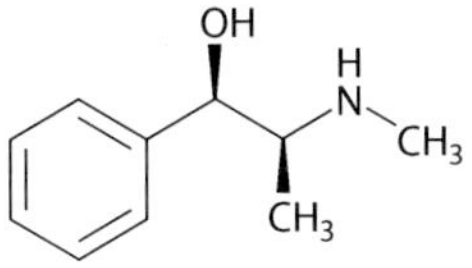

ephedrine

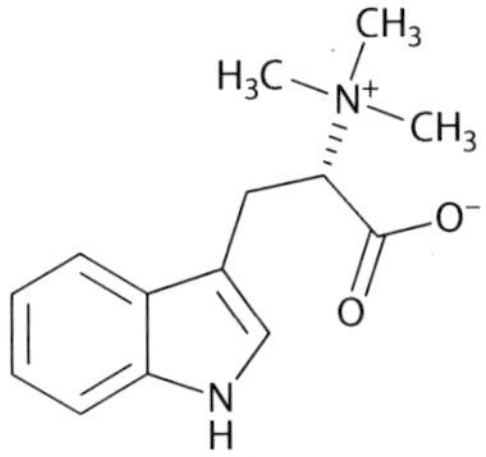

hypaphorine

SPIRAEA TOMENTOSA LINNAEUS

Phylum: Magnoliophyta
Class: Magnoliopsida
Order: Rosales
Family: Rosaceae
Common name: Steeplebush, hardhack
Distribution: North and Central Europe, Eastern North America
Habitat: Wetlands, swamps, pastures, meadows, bogs, lakes, and river margins
Description: This species, which grows to 1.3 m, has an erect and unbranched stem. The leaves (8 × 3 cm) are alternately arranged and are elliptic-oblong to ovate-oblong. While the upper surface of the leaves is dark green, their lower side is pale. The central stem terminates into a dense panicle of pink flowers (20 × 7 cm) with pink petals.
Compounds and activities: The methanol extracts of seeds of this species have shown antimicrobial activities against *Staphylococcus aureus*, *Pseudomonas aeruginosa*, *Escherichia coli*, and *Candida albicans*. The leaves, flowers, and stem of this plant are used in the preparation of medicinal tea for treating dysentery, pregnancy sickness, and easing child birth. The antioxidant activity of the seeds of this species has been tested using trolox extract (TE), and the value recorded was 141,430 μm TE (Borchardt et al., 2009).

NYMPHOIDES CRISTATUM (ROXB.) O. KUNTZE

Phylum: Magnoliophyta
Class: Magnoliopsida
Order: Asterales
Family: Menyanthaceae
Common name: Crested floating heart
Distribution: Asia, China
Habitat: Canals, lakes, ponds
Description: This species has floating stems originated from a buried root stock. Its roots are slender and tuberous. Heart-shaped leaves (5 cm diameter) have smooth margins. There are numerous small white flowers (3 cm wide) in which the petal lobes are crested.
Compounds and activities: The ethyl acetate extract of this species has yielded a triterpenoid compound, viz., bet-20(29)-en-3-ol-28-oic acid, which possesses antibacterial activities against pathogenic bacteria, viz., *Bacillus cereus*, *B. subtilis*, *E. coli*, and *Shigella sonnei*. The values of minimum inhibitory concentration (MIC) of the said compound against these bacterial species were found to be 64, 128, 64, and 32 µg/ml, respectively. This compound also showed prominent cytotoxic activity and the LC_{50} (median lethal concentration) was 15.39 mg/ml (Rahman et al., 2002). Antipyretic and hepatoprotective activities have also been recorded with this compound.

NYMPHAEA LOTUS LINNAEUS

Phylum: Magnoliophyta Class: Magnoliopsida
Order: Nymphaeles Family: Nymphaeceae
Common name: Tiger lotus, white lotus
Distribution: East Africa, Southeast Asia
Habitat: Clear, warm, still waters
Description: This is a perennial plant growing to a maximum height of 50 cm. While its pads float on water, the blossoms rise above water. The flowers are either white or tinged with pink.
Compounds and activities: The bioactive compounds isolated from the leaf extract of this species include tannins, terpenes, flavonoids, alkaloids, anthroquinones, saponins, cardiac glucosides, and phenolics. These antibacterial compounds act against the growth of methicillin and vancomycin-resistant *Staphylococcus aureus* (MRSA and VRSA). The extracts of this species have also been widely used for treating fever and urethral discharge in traditional medicines. The minimum inhibitory concentration (MIC) and minimum bacterial concentration (MBC) ranges are from 5 to 15 and 10 to 30 for MRSA and VRSA, respectively (Akinjogunla et al., 2010).

NYMPHAEA VARIEGATA (DURAND) G.S.MILL (= *NUPHAR VARIEGATA*)

Phylum: Magnoliophyta
Class: Magnoliopsida
Order: Nymphaeles
Family: Nymphaeceae
Common name: Bullhead pond lily
Distribution: North America
Habitat: Ponds, lakes, slow streams, ditches
Description: This dicot, floating, perennial plant has a large spongy rhizome. The floating leaves (5–15 cm across) are rounded and arrow shaped. Their stalk is flattened and poorly winged. The globe or saucer-shaped, solitary, reddish yellow, eight-part flowers (3–5 cm wide) are seen above water.
Compounds and activities: The biologically active two gallotannins and two ellagitannins isolated from this species possess antibacterial properties (Yang et al., 1999). Ellagitannins have also shown antitumor properties.

NYMPHAEA TUBEROSA PAINE (= *NYMPHAEA ODORATA*)

Phylum: Magnoliophyta Class: Magnolipsida
Order: Nymphaeles Family: Nymphaeceae
Common name: White water lily
Distribution: United States, Canada
Habitat: Ponds, lakes, rivers, slow-moving streams
Description: This perennial herb (30 cm height) has flat, rounded, reddish, and alternately arranged floating leaves. The fragrant, white or pink flowers also float on the surface of water. The fruits of this species are green in color.
Compounds and activities: The extracts of this species have been reported to possess antimicrobial and anticoagulant activities. While the ethyl gallate extracts of this species inhibit the growth of *Mycobacterium smegmatis*, its ethanol extracts inhibit the growth of *S. aureus* (Abul-Hajj et al., 1972).

NELUMBO LUTEA WILLD

Phylum: Magnoliophyta Class: Magnoliopsida
Order: Proteales Family: Nelumbonaceae
Common name: American lotus, yellow or water chinquapin
Distribution: North America, Caribbean Islands
Habitat: Lakes, swamps
Description: This species, which grows to a height of 1.5 m, has its roots anchored in mud. The leaves and flowers lie above the water's surface. The leaves measure 30–50 cm in diameter, and the white to pale yellow flowers are 20–30 cm and with 25 petals each.
Compounds and activities: The bioactive alkaloid compounds, such as N-methylasimilobine, anonaine, roemerine, nuciferine, armepavine, N-nornuciferine, and N-norarmepavine, possess antimicrobial activities (Zelenski, 1977).

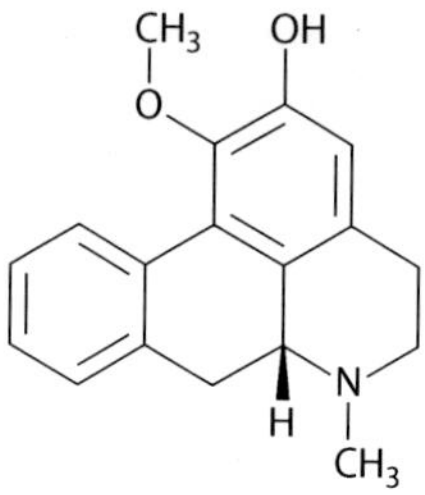

N-methylasimilobine

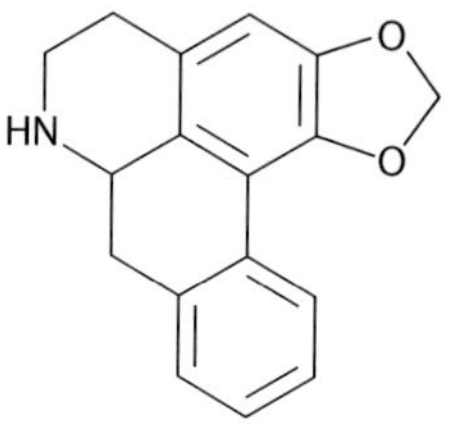

anonaine

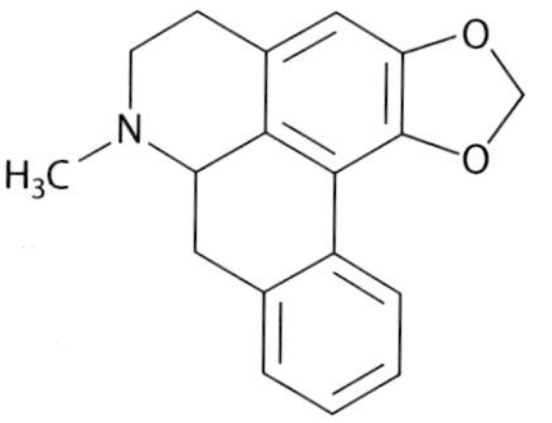

roemerine

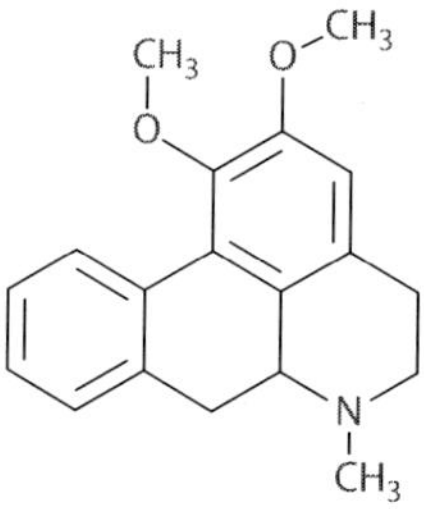

nuciferine

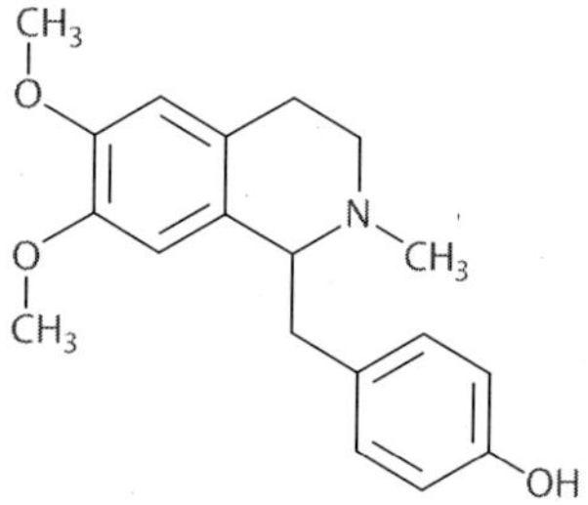

armepavine

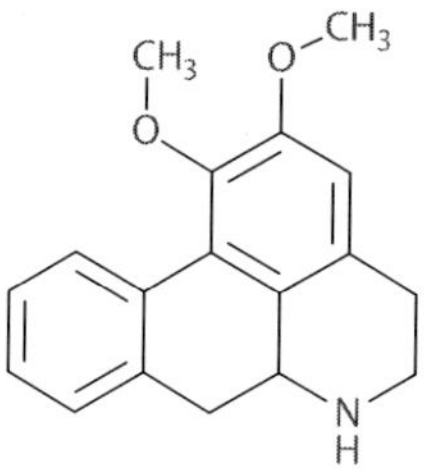

N-nornuciferine

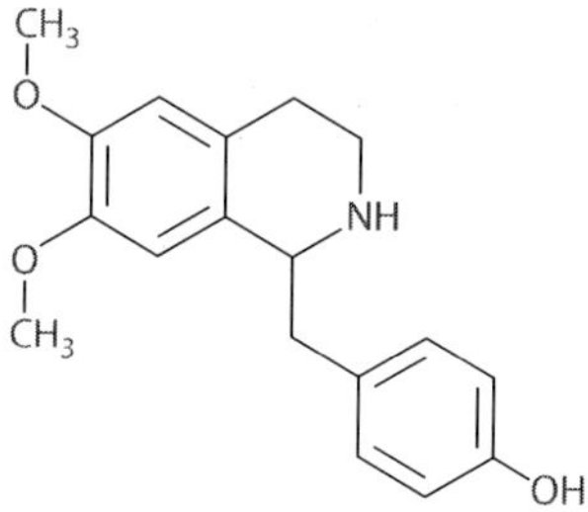

N-norarmepavine

HYGROPHILA SPINOSA (K. SCHUM) HEINE (= *HYGROPHILA AURICULATA*)

Phylum: Magnoliophyta Class: Magnoliopsida
Order: Lamiales Family: Acanthaceae
Common name: Marsh barbel, starthorn
Distribution: India, Bangladesh, Sri Lanka, Burma, Malaysia, Nepal
Habitat: Wetlands, along banks of freshwater and stagnant ditches
Description: This is a perennial, subquadrangular, and stout herb that has an unbranched erect stem of 1.5 m height. Leaves are oblong, lanceolate, or oblanceolate. Purple-blue flowers of 3.5 cm long are present in a whorl of eight at each node.
Compounds and activities: The alkaloids of this species have also shown antibacterial activities against Gram-positive bacteria, viz., *Bacillus subtilis*, *B. cereus*, and *Staphylococcus aureus*, and Gram-negative *Proteus vulgaris*. The MIC value of stem methanol extract and flower ethyl acetate extract of this species against *Bacillus subtilis* was 0.5 mg/ml, and MIC values of its flower and leaf extracts against *Proteus vulgaris* were 0.031 and 0.25 mg/ml, respectively (Jani et al., 2012). This plant is also used in traditional medicines for treating cough, pain, inflammation, urinary infection, edema, gout, and urethral discharge. The bioactive compounds isolated from the stem and leaves of this species include saponins, alkaloids, steroids, tannins, flavonoids, and triterpenoids. These compounds possess antipyretic, antidiabetic, antinociceptive, antitumor, anti-inflammatory, antioxidant, analgesic, hepatoprotective, hypoglycemic, hematinic (hematopoietic), diuretic, free radical scavenging, androgenic, anabolic, and antihelmintic activities (Kshirsagar et al., 2010).

CERATOPHYLLUM DEMERSUM LINNAEUS

Phylum: Magnoliophyta
Class: Magnoliopsida
Order: Ceratophyllales
Family: Ceratophyllaceae
Common name: Hornwort, coon tail
Distribution: North America, cosmopolitan in temperate regions
Habitat: Lakes, ponds, quiet streams, still and slow-moving waters
Description: The stem of this submerged and free-floating species measures 3 m (height), and its leaves are in whorls of 6–10. The leaf (40 mm long) is simple and forked into 3–10 thread-like segments. This species is monoecious, and the flowers are very small (2 mm long). The fruit is a small nut of only 5 mm long.
Compounds and activities: The bioactive antibacterial compounds isolated from this species include alkaloids and lipophilic flavonoids. The antibacterial polyphenols isolated from this species are known to act against enteric bacteria. Further, the acetone extracts of this plant have shown antibacterial properties against *Bacillus* sp. and *P. aeruginosa* (Ghobrial et al., 2007).

CAMPTOTHECA ACUMINATA DECNE

Phylum: Magnoliophyta Class: Magnoliopsida
Order: Cornales Family: Cornaceae
Common name: Happy tree
Distribution: China, Tibet
Habitat: Humid streams and river banks
Description: This is an endangered species growing to a maximum height of 25 m. The leaves are reddish green and leathery. The whitish flowers of this species blossom annually.
Compounds and activities: A total of 26 endophytic fungi were isolated from the stem, leaves, and fruits of this species. Fermentation broths from most of the isolates showed antifungal activity and 50% exhibited antibacterial activity. The strongest antimicrobial activity was exhibited by strains XSY 10 and XSY 15 against *Rhizoctonia solani* and *Fusarium oxysporum* with 75 and 67% inhibition, respectively (Ding et al., 2010). A pentacyclic quinoline alkaloid compound, camptothecin, which inhibits the growth of cells relating to tumor and leukemia, has been isolated from this species. The synthesized forms of camptothecin are topotectan and irrinotectan. The extracts of this species also have other bioactive compounds to treat psoriasis, cold, and liver and stomach ailments.

POLYGONUM HYDROPIPER (LINNAEUS) (= *PERSICONA HYDROPIPER*)

Phylum: Magnoliophyta Class: Magnoliopsida
Order: Polygonales Family: Polygonaceae
Common name: Marsh pepper, knotweed, water pepper
Distribution: Australia, New Zealand, Asia, Europe, North America
Habitat: Damp places, shallow water
Description: The species is erect ascending or decumbent. The stem, which is glabrous or slightly pubescent, is green in the beginning and red when aged. Leaves (5 × 2 cm) are glabrous and lanceolate. Inflorescences are lax

flowered, often with a small leaf or bran at the base of the terminal spike. Small inflorescences arise at all the nodes producing cleistogamous flowers.

Compounds and activities: Bicycle pesquiterpenoids, viz., polygodial (tadeonal, an unsaturated dialdehyde with a drimane backbone) and waburganol and rutin, have been isolated from plant extracts. Its essential oil contains both monoterpenoids and sesquiterpenoids, viz., α-pinene, 1,4-cineol, fencnone, α-humulene, β-caryophyllene, trans-β-bergamotene, and carboxylic acids. The chloroform root extract of this species showed antibacterial activities against four Gram-positive bacteria, *B. subtilis*, *B. megaterium*, *S. aureus*, and *Enterobacter aerogenes*, and MIC values against these species were found to range from 16 to 64 μg/ml. The antifungal activities were also found strong with six fungal species, *A. fumigatus*, *A. niger*, *A. flavus*, *C. albicans*, *Riopus oryzae*, and *Trichophyton rubrum*. Values of 20 and 19 mm diameter of zone of inhibition were observed against *A. fumigates* and *A. niger*, respectively, at a concentration of 300 μg/disc (Hasan et al., 2009).

RUMEX JAPONICUS HOUTTUYU

Phylum: Magnoliophyta
Class: Magnoliopsida
Order: Caryophyllales
Family: Polygonaceae
Common name: Gishigishi
Distribution: Japan, Russia, Korea, China
Habitat: River side, wet meadows, ditches, stream banks, wet valleys

Description: This perennial species grows to a maximum height of 100 cm, and its leaves are 10–25 cm long. The flowers are green in color.

Compounds and activities: The aerial and root extracts of this species have been reported to contain antioxidant flavonoids and anthroquinones such as emodin, chrysophenol, and physcion. The ethyl acetate extract of this species also possesses the strongest antibacterial activity against *B. subtilis*, *B. cereus*, and *E. coli* (Elzaawely et al., 2005).

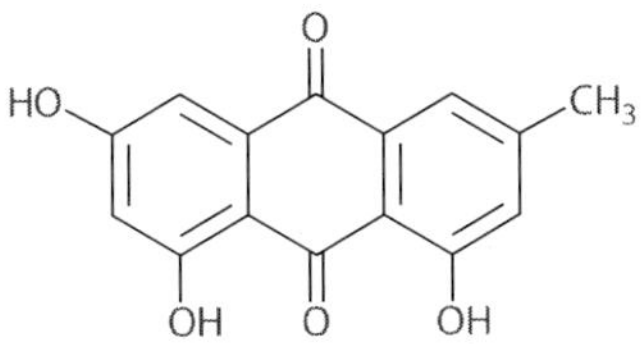

emodin

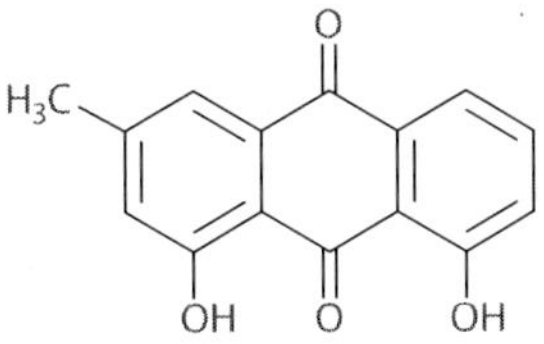

chrysophanol

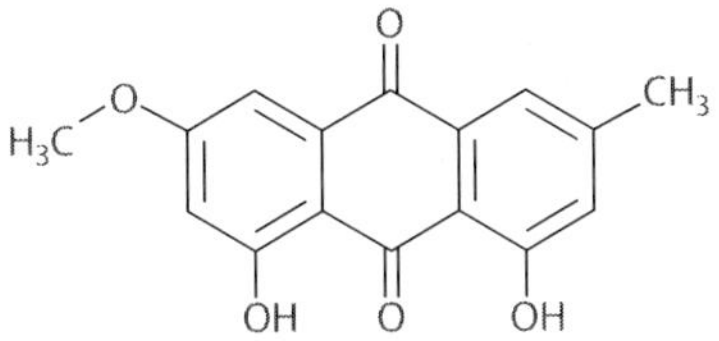

physcion

AMORPHA FRUTICOSA LINNAEUS

Phylum: Magnoliophyta
Class: Magnoliopsida
Order: Fabales
Family: Fabaceae
Common name: False indigo bush
Distribution: United States, Canada, Mexico, Europe, Asia
Habitat: Freshwater lakes, streams
Description: This is a perennial shrub growing to a maximum height of 6 m. Foliage is green, and the leaves are made up of many hairy, oval-shaped, spine-tipped leaflets. Inflorescence is a spike-shaped raceme of many purple flowers.
Compounds and activities: New antimicrobial bibenzyl metabolites, viz., amorfrutin A and B have been isolated from the wax fractions of the fruit of this species (Mitscher et al., 1981). The other compound isolated from this species is 6-O-β-D-glucopyranosyl-12-a-hydroxy dalpanol, a rotenoid.

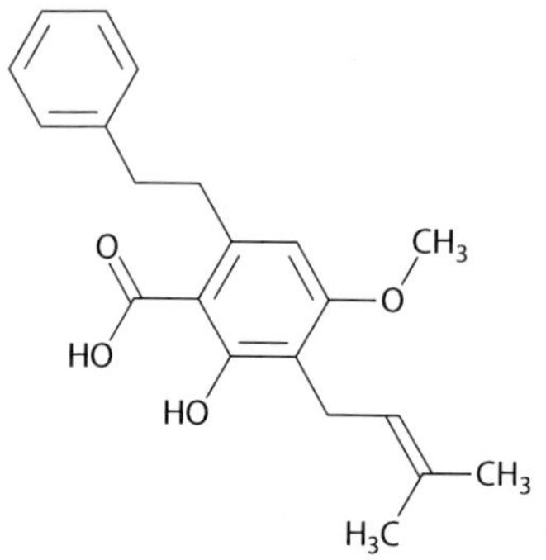

amorfrutin A, B

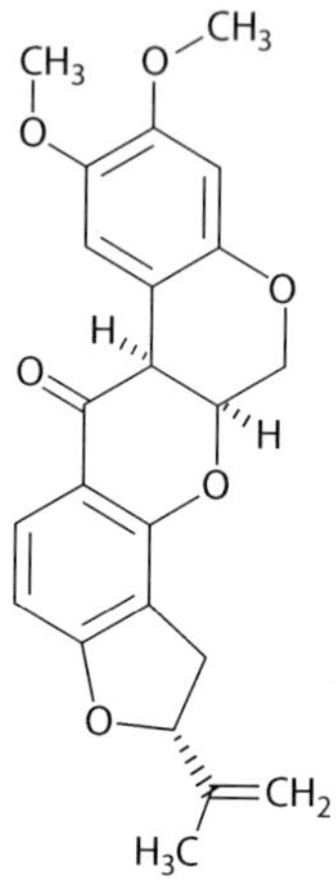

rotenoid

LEMNA GIBBA LINNAEUS

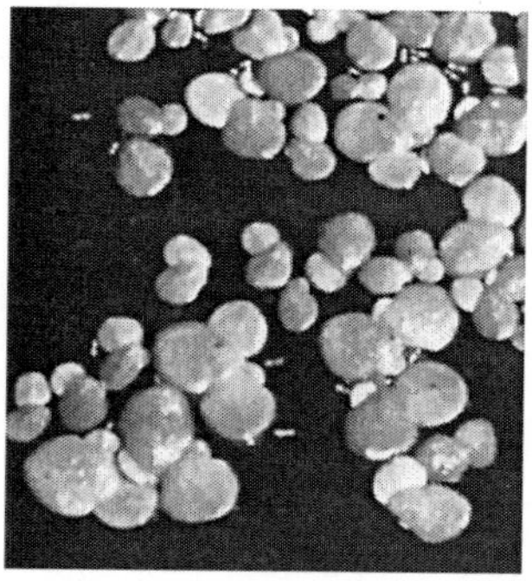

Phylum: Magnoliophyta
Class: Liliopsida
Order: Arales
Family: Lemnaceae
Common name: Common duckweed, fat duckweed, swollen duckweed
Distribution: Throughout world (Europe, Southwest Asia, Japan, India, Sri Lanka, Northeast and South Africa, South America)
Habitat: Freshwater ponds, marshes, quiet streams
Description: This floating species, which forms green mats, has a thallus of 5 mm diameter. A single root hangs down from the thallus into the water.
Compounds and activities: The methanol extracts of this species have insulin-like products that are known to act against enteric bacteria (Ghobrial et al., 2007).

ECHINODORUS MACROPHYLLUS (KUNTH) MICHELI, 1881

Phylum: Magnoliophyta
Class: Liliopsida
Order: Alismatales
Family: Alismataceae
Common name: Leather hat
Distribution: Guyana, Brazil, Argentina
Habitat: Freshwater habitats, rivers
Description: The stem of this plant is upright and cylindrical. The leaves (30 × 30 cm) are membranous and triangularly ovate. The inflorescence, which is rarely racemose, is paniculate with 6–14 whorls containing 6–10 flowers each.
Compounds and activities: From the leaf extract of this species, bioactive compounds, viz., echinophyllins A, B, C–F (N_2-containing clerodane diterpenoids) and chapecoderins A–C (labdane-derived diterpenoids), have been isolated. Among them, echinophyllins have been reported to possess antimicrobial and cytotoxic properties (Kobayashi et al., 2000a,b). This species has also been used in traditional medicines for treating inflammation, rheumatic diseases, and kidney dysfunctions.

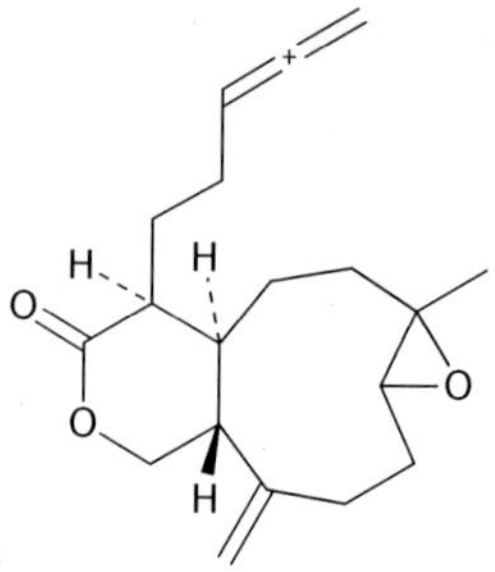

89

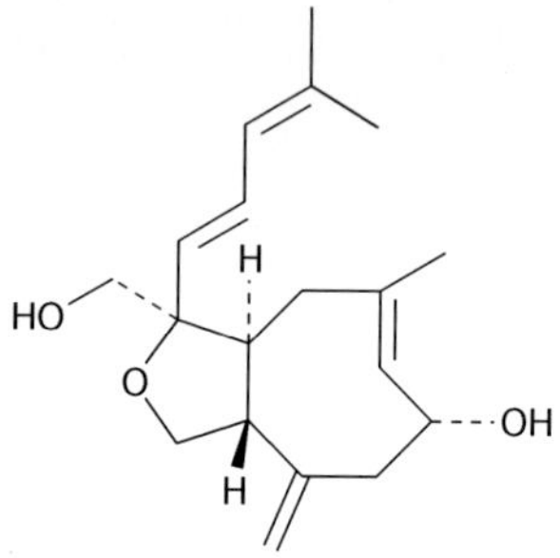

90

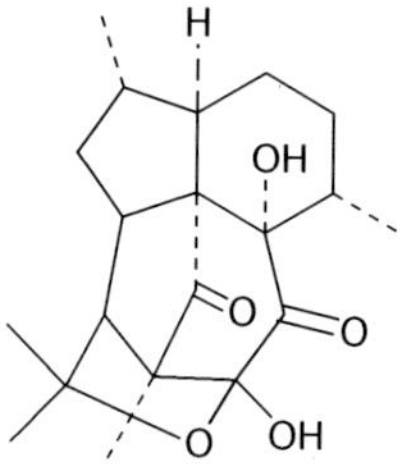

91

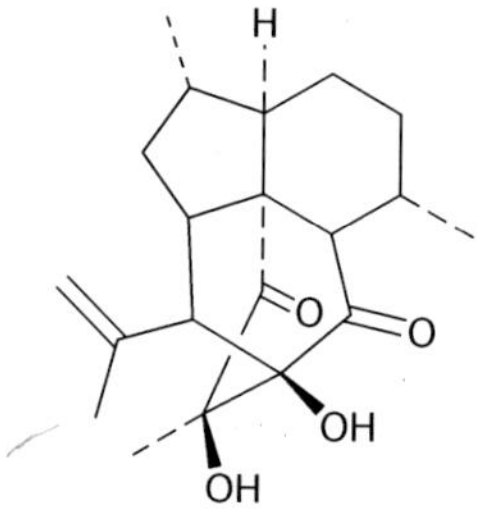

92

echinophyllins C-F

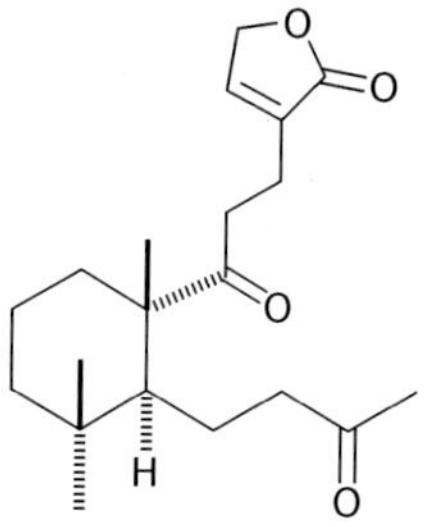

chapecoderin A

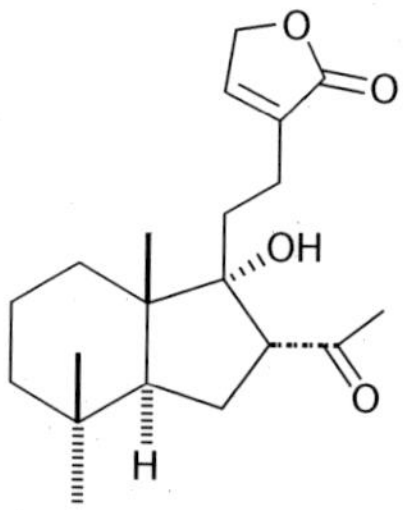

chapecoderin B

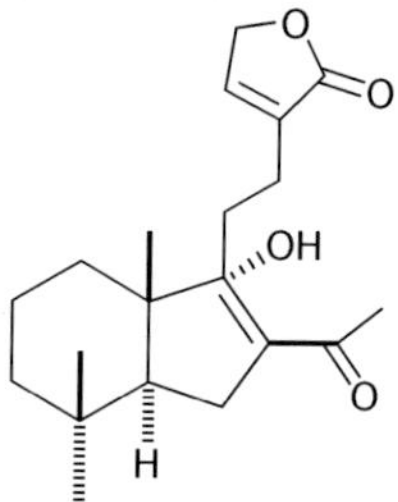

chapecoderin C

chapecoderins A-C

SAGITTARIA SAGITTIFOLIA LINNAEUS

Phylum: Magnoliophyta
Class: Liliopsida
Order: Alismatales
Family: Alismataceae
Common name: Arrowhead
Distribution: Europe, Asia
Habitat: Wetlands
Description: It is a herbaceous perennial plant growing in shallow waters (50 cm deep). Leaves above water are arrow shaped, and their blades measure 25 cm long and 20 cm wide. Narrow, linear, submerged leaves are also present. Flowers that are 2 cm broad have white petals and three small sepals.
Compounds and activities: The ent-rosane diterpenoids, viz., sagittines A–D and 13-epi-manoyl oxide-19-O-α-L-2-5′-diacetoxyrabine furonoside, have been isolated from this species. Among them, sagittines A–D exhibited antibacterial activity against oral pathogens such as *Streptococcus mutans* and *Actinomyces naeslundii*. The MIC values of these species ranged from 62.5 and 125 μg/ml (Liu et al., 2006).

NAJAS MARINA LINNAEUS

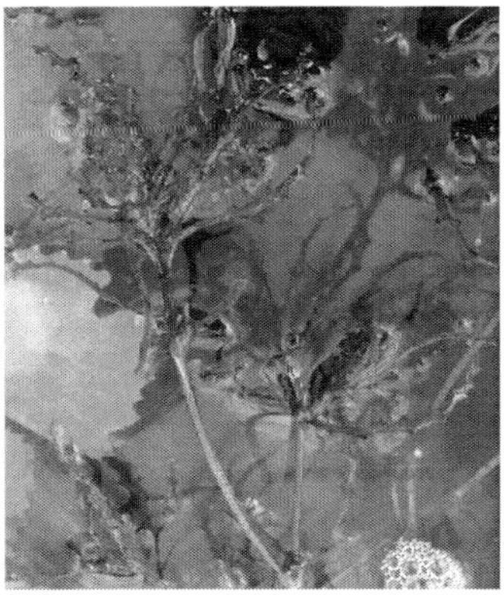

Phylum: Magnoliophyta Class: Liliopsida
Order: Alismatales Family: Najadaceae
Common name: Holly-leaved naiad, spiny naiad
Distribution: Worldwide distribution
Habitat: Freshwater and brackish water ponds
Description: This is a perennial species growing to a maximum height of 50 cm. The stem is slender and branched. The tough, narrow, and evenly spaced leaves (4 cm × 3 mm) are edged with very small saw-like teeth. Many dioecious flowers are present in the leaf axis.
Compounds and activities: The methanol extracts of the whole plant showed antibacterial activities against *Staphylococcus aureus*, and the MIC value was found to be 0.031 mg/ml (Jani et al., 2012).

NAJAS GRAMINEA DELITE

Phylum: Magnoliophyta Class: Liliopsida
Order: Alismatales Family: Najadaceae
Common name: Rice field water-nymph
Distribution: Temperate areas of world: Southern Europe to Eastern Asia
Habitat: Ponds, rice fields, still or slow-moving waters
Description: This species grows to a maximum height of 30 cm, and its flowers are monoecious; i.e., both male and female flowers are in the same plant.
Compounds and activities: Ninety-five percent ethanolic, the extract of this species showed antibacterial properties against Gram-positive bacteria such as *Bacillus cereus* and *Staphylococcus aureus*, and the values of diameter of inhibition zone (mm ± SD) in these bacterial species were 0.20 ± 0.45 and 0.32 ± 0.46, respectively (Hamkrasri, 2011). Its methanolic and ethanolic extracts have also shown high antioxidant (scavenging) activities.

STUCKENIA PECTINATA (LINNAEUS) BOERNER (= *POTAMOGETON PECTINATUS*)

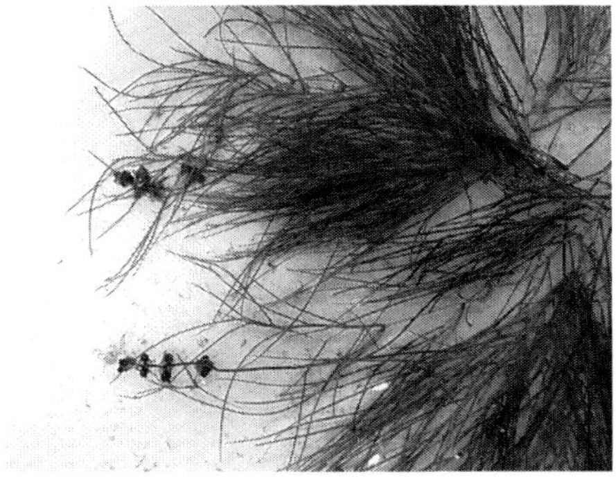

Phylum: Magnoliophyta Class: Liliopsida
Order: Alismatales Family: Potamogetanaceae
Common name: Sago pond weed, ribbon pond weed, fennel pond weed
Distribution: Cosmopolitan, all continents except Antarctica
Habitat: Freshwater and brackish water habitats
Description: This fully submersed plant species has no floating or submersed leaves. The flowers are pollinated by winds, and reproduction is by vegetative method with tubers and sexually by seeds.
Compounds and activities: This species has yielded ent-labdane diterpenes, viz., methyl-15,16-epoxy-12(R)-acetoxy-8,13,14-ent-labdatrien-19-oate, 15,16-epoxy-12(R)-acetoxy-8, and 13,14-ent-labdatrient-19-oic, which possess antibacterial activities and act against mainly enteric bacteria (Ghobrial et al., 2007).

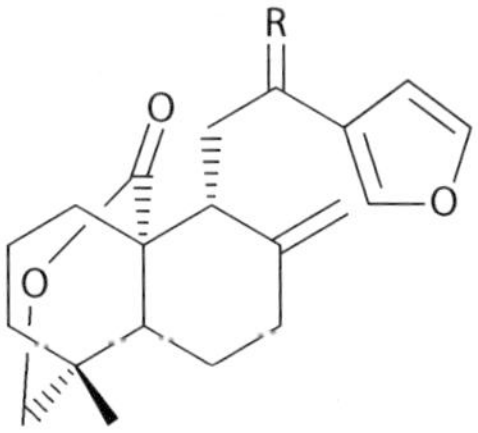

ent-labdane diterpenes

POTAMOGETON NATANS LINNAEUS

Phylum: Magnoliophyta Class: Liliopsida
Order: Alismatales Family: Potamogetanaceae
Common name: Broad-leaved pond weed, floating pond weed
Distribution: Northern hemisphere, including Britain
Habitat: Lakes, ponds, ditches, quiet or slow-flowing freshwaters
Description: The stems (1–2 m) of this plant species are cylindrical without many branches. Both floating and submerged leaves are present in the same plant. The floating dark green leaves (5–10 cm long) are ovate to oblong, and ovate and cordate at the base. They are leathery, opaque, and rounded at the base and pointed at the tips. The submerged leaves are grass-like phyllodes, which are modified leaf stalks. The hermaphrodite flower spikes (5–10 cm long) are dense and cylindrical.
Compounds and activities: The bioactive compounds isolated from this species are potamogetonin, a new furanoid diterpene, and seven furanto-ent-labdane diterpenes (Greca et al., 2001). While the synthetic potamogetotonin analog has antimycobacterial properties, the diterpenes show antibiotic (antialgal) activities.

POTAMOGETON NODOSUS POIR

Phylum: Magnoliophyta Class: Liliopsida
Order: Alismatales Family: Potamogetonaceae
Common name: Long-leaf pond weed
Distribution: Europe, Asia, Americas
Habitat: Ponds, lakes, ditches, streams
Description: This monocot, perennial herb has thin branching stems of 1 m length. The floating and submerged leaves (15 × 4 cm) are linear to lance shaped. The inflorescence is a spike of many small flowers, which arise from the water on a peduncle.
Compounds and activities: From the ethanol extract of this species, bioactive compounds such as 15,16-epoxy-12-oxo-8,13,14-labdatrien-20,19-olide and 2-hydroxyheptane-3,5-dione have been isolated. These compounds show antibacterial (against both Gram-positive and Gram-negative bacteria), cytotoxic, and antiviral activities. Other compounds isolated from this species include potamogeton nodosus lectins (PNLs) 1–6, which possess cytotoxic, anticancer, and antitumor properties (Pervin et al., 2006).

VETIVERIA ZIZANIOIDES (LINNAEUS) NASH

Phylum: Magnoliophyta Class: Liliopsida
Order: Poales Family: Poaceae
Common name: Khas grass
Distribution: India, Bangladesh, Myanmar
Habitat: Swamps
Description: It is a perennial grass with thick fibrous adventitious roots that are aromatic and highly valued. This densely tufted grass, which has a maximum height of 2 m, arises from an aromatic rhizome. Leaves are narrow, erect, and keeled with scabrid margins. Inflorescence is a panicle up to 50 cm long of numerous slender racemes in whorls on a central axis.
Compounds and activities: The bioactive fractions of the essential oil of this species, viz., V_z-2, V_z-7, V_z-8, and V_z9, on purification yielded six compounds, viz., 5,10-pentadecadiyn-1-ol, α-curcumene, hydroxy

junipene, (+)-cycloisosativene, valencine, and selino-3,7(11)-diene. All these compounds showed significant antimycobacterial activity against the drug-resistant strains (MDR-R and MDR-40) of *Mycobacterium smegmatis*, and their MIC was in the range of 31.25–62.5 μg/ml (Jeyashree et al., 2011). The crude oil of this species also yielded bioactive compounds, viz., β-vetivenene, β-vetivone, and α-vetivone, which possess antioxidant properties (Kim, 2005).

TYPHA CAPENSIS (ROHRB.) N.E. BR.

Phylum: Magnoliophyta
Class: Liliopsida
Order: Poales
Family: Typhaceae
Common name: Bulrush
Distribution: Throughout world
Habitat: Marshes, stream banks, dams, lakes
Description: This is a perennial leafy herb. The stem is erect and simple and terminates in dense cylindrical velvety-brown flower spikes. Leaves are long, bluish gray to green, and strap shaped with a maximum length of 1.5 m. Flowers are minute and unisexual. Male and female flowers occupy the top and lower portions of the spike, respectively.
Compounds and activities: Methanol extracts of dried ground leaves and rhizomes of this species showed significant antibacterial activities. Methanol was found to be the best extractant with an average MIC value of 0.75 mg/ml (rhizome) and 0.21 mg/ml (leaves). *E. coli* and *Enterococcus faecalis* were the most sensitive, with average MIC values of 0.53 and 0.42 mg/ml, respectively. The MIC values for the positive control (ampicillin) ranged from 0.08 to 0.16 mg/ml (Masoko et al., 2008).

EICHHORNIA CRASSIPES (MART) SOLMS

Phylum: Magnoliophyta
Class: Liliopsida
Order: Commelinales
Family: Pontederiaceae
Common name: Water hyacinth
Distribution: Africa, New Zealand
Habitat: Freshwater habitats, brackish waters (less than 15%)
Description: This floating plant has rosettes of rounded, leathery, glossy, waxy, green leaves and feathery dark roots. The inflorescence is an aerial spike of 30 cm diameter.
Compounds and activities: The crude methanolic extract of this species containing an alkaloid and four phthalate derivatives exhibited 21% antimicrobial activities against *Bacillus subtilis*, *Escherichia coli*, *Streptococcus aureus*, *S. faecalis* (after 24 h), and *Candida albicans* (after 72 h). The values of diameter of inhibition zone (mm) for these microbial species were 12 ± 0.5, 12 ± 0.0, 12 ± 0.2, 14 ± 0.2, and 13 ± 0.5, respectively (Shanab et al., 2010). The bioactive compounds isolated from this species include phenylphenalconones (phenyl propanoid derivatives), methyl derivatives of phenylnaphthalene dicarboxylic acid, and phenylnaphthalene carboxylic acid (Greca et al., 2009). The benzoindenone of this species possesses antimicrobial and antioxidative properties. The sulfated polysaccharides, viz., heterogalactans of the roots and rhizomes of this species, possess anticoagulant properties by prolonging the coagulation time of blood. Further, the extracts of this species have also been reported to possess antiproliferative, antitumoral, anticomplementary, and anti-inflammatory properties.

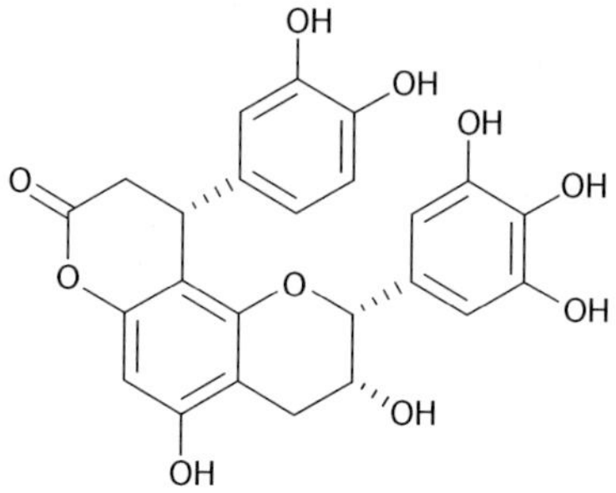

phenylphenalconones (phenyl propanoid derived)

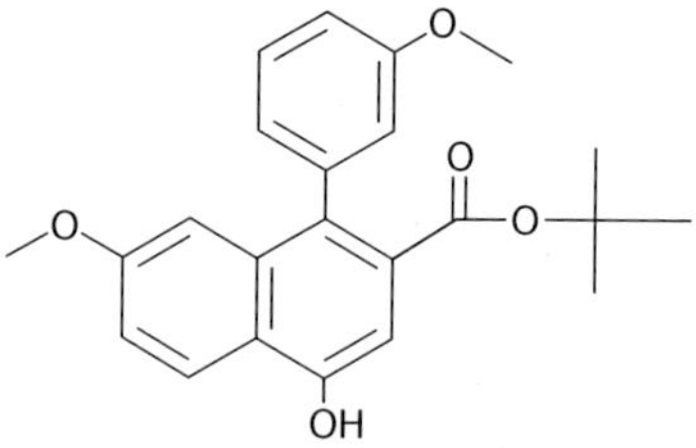

methyl derivatives of phenylnaphthalene dicarboxylic acid

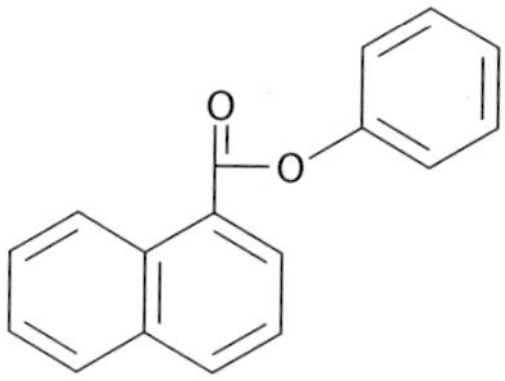

phenylnaphthalene carboxylic acid

MONOCHORIA VAGINALIS (BURM. F.) C. PRESL EX KUNTH

Phylum: Magnoliophyta Class: Liliopsida
Order: Commelinales Family: Pontederiaceae
Common name: Heartleaf base
Distribution: Asia, Pacific Islands
Habitat: Rice fields, freshwater bodies
Description: This is a perennial or annual herb growing in water from a small rhizome. Green leaves are shiny with a maximum length of 12 cm and width of 10 cm. Inflorescence bears 5 to 25 flowers. Each flower has six purple-blue tepals of 1 cm long each.
Compounds and activities: Two new bioactive compounds, viz., (IOZ)-1-(2,6-dihydroxyphenyl) octadec-10-ene-1-one and (20R,24R)-campest-5-ene-3β,4β-diol, together with nine known compounds, have been isolated from the whole plant of this species (Row et al., 2004). Among them, the first two compounds showed antibiotic properties.

COMMELINA BENGHALENSIS LINNAEUS

Phylum: Magnoliophyta Class: Liliopsida
Order: Commelinales Family: Commelinaceae
Common name: Day flower
Distribution: Asia, Africa, West Indies, North America, China
Habitat: Rice fields, ditches, moist soils
Description: It is a perennial herb. Stems are branched and some of them grow underground and bear reduced leaves. Leaves are ovate-elliptic or oblong. Flowers are in one or two cyme branches and are bluish violet.
Compounds and activities: The ethanol extracts of this species, containing bioactive compounds such as alkaloids, lactones, coumarins, triterpenoids, phenols, tannins, quinones, and saponins, were found to be active against *Candida albicans*, *E. coli*, and *S. aureus*. The values of diameter of zone of growth inhibition in the absolute alcohol extract of the dried plant of this species for the above microbial species were 7, 4, and 6 mm, respectively (Armando and Dennis, 2010). Further, the above compounds also possess anti-inflammatory and abortifacient properties.

HYDRILLA VERTICILLATA (L.F.) ROYLE

Phylum: Magnoliophyta
Class: Liliopsida
Order: Hydrocharitales
Family: Hydrocharitaceae
Common name: Hydrilla, water thyme
Distribution: Asia, North America, South America, Europe, Africa, Australia, New Zealand
Habitat: Freshwater systems
Description: This submerged macrophyte is a polymorphic and monoecious or dioecious species and has many horizontal aboveground stems (stolons) and underground stems (rhizomes). The leaves (20 × 4 mm) occur in whorls of four to eight. The female flowers (50 × 8 mm) have three whitish sepals and three translucent petals.
Compounds and activities: The compounds isolated from the ethanol and methanol extracts of this species are loliolide, thymidine, and sulfur gamma, which possess antibacterial and antioxidative properties (Pal et al., 2004). Its otteliones A and B exhibit potent antitumor activity (Araki et al., 2003). This species is also a potential source of calcium and vitamin B12 and is rich in lysine, minerals, other vitamins, chlorophyll, proteins, and essential fatty acids. Hence, it has been reported to support immune systems of humans besides taking care of weight loss, healthy skin, and strong bones.

PHALARIS ARUNDINACEA LINNAEUS (1753)

Phylum: Magnoliophyta Class: Liliopsida
Order: Cyperales Family: Ponaceae
Common name: Reed canary-grass, ribbon grass
Distribution: North America, Europe
Habitat: Wetlands, margins of lakes, brooks, and swamps, roadside ditches, river dikes, shallow marshes, meadows
Description: This rhizomatous perennial plant grows to a maximum height of 2 m. The stems (1 cm diameter) are sturdy and hollow. The leaf blades (2 cm diameter) are flat and hairless. The flowers are borne in panicles (15 cm long) on culms high above leaves.
Compounds and activities: The leaf extracts of this species have bioactive compounds such as the central nervous system (CNS)-active natural products, viz., N,N-dimethyltryptamine (DMT) and 5-methoxy dimethyl tryptamine (5-MeO-DMT), and β-carbolines and hordenine. The compound hordenine possesses antibacterial and antibiotic activities. β-Carbolines act as monoamine oxidase inhibitors (MAOIs) (antidepressants). DMT is considered to be a hallucinogenic, enthenogenic, or psychedelic drug (mind-expanding drug), and it functions as a trace amine neurotransmitter. 5-MeO-DMT has also been reported to be used as an enthenogen (psychoactive substance) (McClatchey et al., 2009).

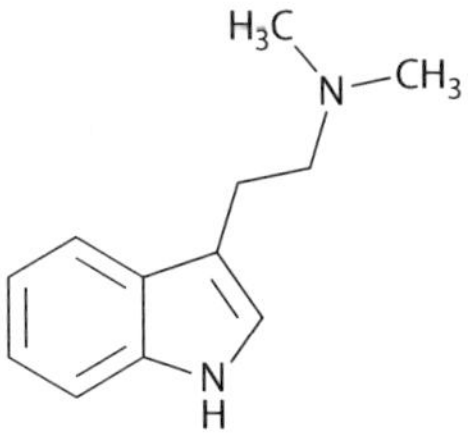

DMT (dimethyltryptamine)

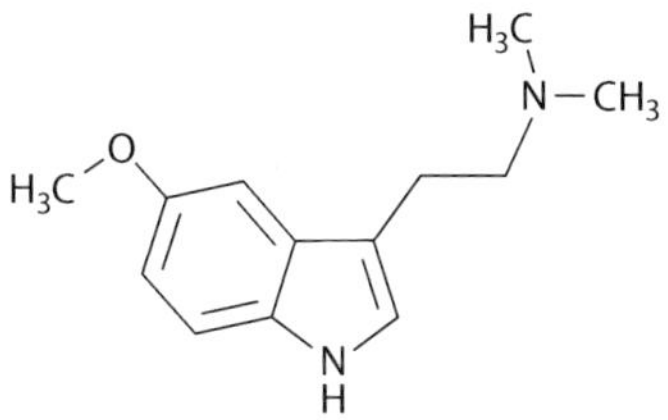

5 methoxy dimethyl tryptamine (5-MeO-DMT)

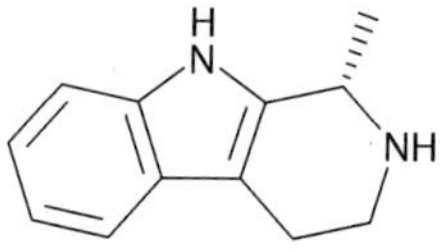
beta carbolines

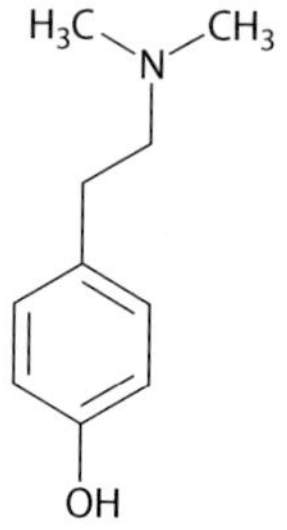
hordenine

ACORUS CALAMUS LINNAEUS

Phylum: Magnoliophyta
Class: Liliopsida
Order: Acrales
Family: Araceae
Common name: Sweet flag
Distribution: Arabia, India, Russia, China, Japan, Burma, Sri Lanka, Indonesia, Australia, Northern United States, Canada, Europe
Habitat: Semiaquatic, ponds, swamps, marshes, ditches, river edges, lakes
Description: This monocot plant species is known for its scented rhizomes and leaves. The leaves are 1.8 cm wide with undulated margins. The flowers (3–5 mm long) are in a spadix (9 cm long).
Compounds and activities: The compounds isolated from the alcoholic rhizome and leaf extracts of this species include calamusins A-1 (norsesquiterpenes), α- and β-asarone, ursolic acid, (E,Z)-2,4-decadienol, linalool, farnesol, methylleugenol, β-pinene, (E)-caryophyllene, β-elemine, ocimene, and

aromadendrene. The ethyl acetate extracts of rhizomes and leaves of this species showed pronounced antifungal activity with diameter zones of inhibition ranging from 20 to 28 and from 18 to 25 mm, respectively. They also showed antiyeast activity with diameter zones of inhibition ranging from 22 to 25 and from 20 to 23 mm, respectively. The minimum inhibitory concentration (MIC) of rhizome and leaf extracts for antifungal activity measured was 2–4 mg/ml. The MIC value for antibacterial activity, however, was high (16–42 mg/ml). The compounds α- and β-asarones (phenylpropanoids) of this species are said to be responsible for these antimicrobial activities (Asha Devi and Ganjewala, 2009). Asarones and ursolic acid have also been reported to possess antiepileptic and antihelmintic activities. The extracts of roots possess antimalarial properties.

THALIA MULTIFLORA HORKEL EX KOERNICKE

Phylum: Magnoliophyta   Class: Liliopsida
Order: Zingiberales   Family: Marantaceae
Common name: Fire flag
Distribution: Tropical America, tropical Africa, Argentina, Paraguay
Habitat: Freshwater habitats
Description: This species, which grows to a maximum height of 2 m, is known for its large starch rhizome. The blue-green leaves are ovate to lanceolate. Small violet flowers are seen on a panicle (20 cm length) that is held above the foliage.
Compounds and activities: From the aerial parts of this plant, bioactive antitubercular sterols, viz., stigmast-5-en-3β-ol-7-one, stigmast-4-en-6β-ol-3-one, stigmast-5,22-dien-3β-ol-7-one, and stigmast-4,22-dien-6β-ol-3-one, have been isolated. These compounds have shown MIC values of 1.98 ± 0.02, 4.2 ± 0.17, 1.0 ± 0.06, and 2.2 ± 0.3 μg/m, respectively (Lugo et al., 2005).

3 Antiprotozoal Compounds

Protozoan parasites may cause severe diseases like malaria, leishmaniasis, Chagas disease, sleeping sickness, and amoebiasis. As of today, there are no protective vaccines against any of these diseases. Further, parasite resistance to existing drugs has become a serious problem. Freshwater plants offer a variety of antiprotozoal compounds, such as alkaloids, coumarins, flavonoids, terpenoids, phenolics, lignans, tannins, quinines, quassinods, xanthones, etc.

RHIZOCLONIUM HIEROGLYPHICUM (C. AGARDH) KUTZING, 1845

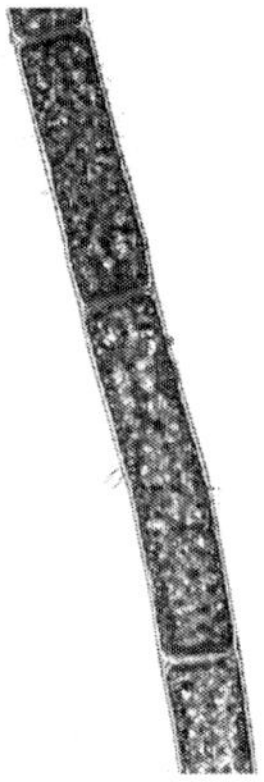

Phylum: Chlorophyta　　Class: Ulvophyceae

Order: Cladophorales　　Family: Cladophoraceae

Distribution: Europe, Central America, Southwest Asia, Central Asia, Australia, New Zealand, Pacific Islands

Habitat: Freshwater, brackish water, and marine habitats

Description: In this unbranched, filamentous species, rhizoids are rare or absent. The cells (250 × 40 µm) of the filaments are thin walled and cylindrical and are with thick and stratified walls occasionally. The parietal chloroplasts are net-like with several pyrenoids.

Compounds and activities: This species has yielded eight fatty acid amides, including (Z)-9-octadecenamide as the major bioactive compound. The latter has shown antimalarial activity with mosquito larvicidal characteristics (Wangchuk, 2004). Further, the cyclic diterpenoids, viz., β-sitosterol (sterol) and transphytol (cyclic diterpenoid), of this species have shown antibacterial activities.

LYNGBYA AESTUARII LIEMAN EX GOMONT, 1892

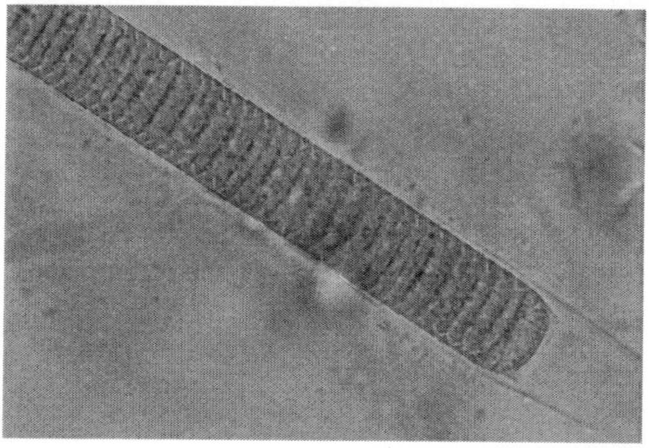

Phylum: Cyanobacteria
Class: Cyanophyceae
Order: Oscillatoriales
Family: Oscillatoriaceae
Distribution: Europe, North America, Central America, South America, Africa, Southwest Asia, Southeast Asia, Australia, New Zealand, Pacific Islands, Indian Ocean Islands
Habitat: Freshwater, brackish water, and marine habitats
Description: The filaments, which are long, curved, or straight, are in clusters or layered masses. Young filaments have thin and smooth sheaths, and in matured filaments, the sheath is thick, wide, lamellated, and yellow-brown. The trichomes (10–20 μm wide), which are cylindrical, are brown-green, olive-green, dark green, or blue-green. The cells are short and discoid with small granules.
Compounds and activities: The bioactive compound isolated from this species is geosmin (an odorous metabolite). The methanolic extracts of this species have shown antileishmanial activity against the protozoan parasite, *Leishmania donovaini*, which causes skin sores, stuffy nose, and breathing disorders in humans. Further, the methanolic crude extracts (100 μg/ml) of this species have shown antibacterial activities against Gram-positive and Gram-negative bacterial species (Kumar et al., 2012) and antifungal properties against *Candida albicans*.

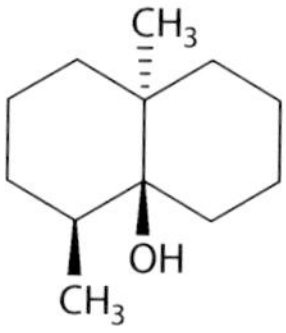

geosmin

NOSTOC MUSCORUM C. AGARDH, 1812

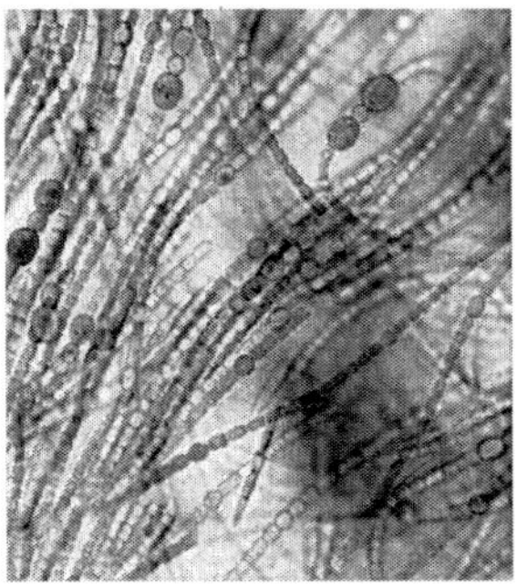

Phylum: Cyanobacteria Class: Cyanophyceae
Order: Nostocales Family: Nostocaceae
Distribution: Europe, Southwest Asia, Australia, New Zealand
Habitat: Freshwater and terrestrial environments
Description: This free-living filamentous species has Gram-negative, green-brown-colored cells that are phototrophic, atmospheric-nitrogen fixers. These organisms form spores under desiccation conditions.
Compounds and activities: The bioactive compounds isolated from this species are poly-β-hydroxybutyrate, nostocyclamide, chlorinated nostocarboline, and β-carbolines. While the chlorinated nostocarboline possesses antiplasmodial (antimalarial) activity, β-carboline alkaloids have played a key role in the pharmacology of indigenous hallucinogenic drugs (Mallick et al., 2010).

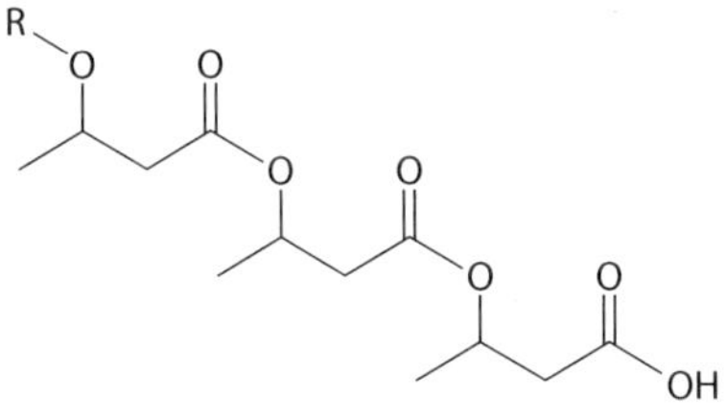

poly-beta-hydroxybutyrate

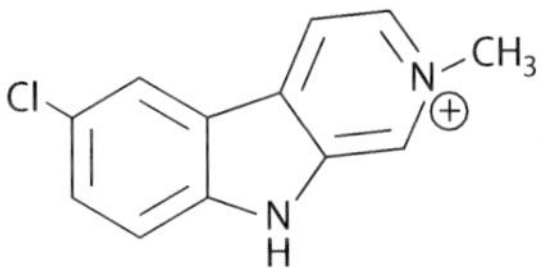

nostocyclamide

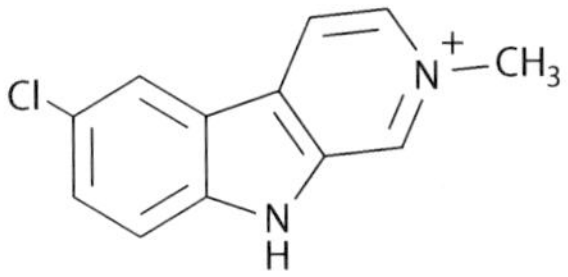

chlorinated nostocarboline

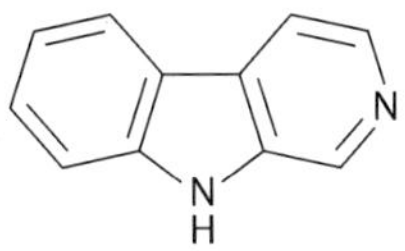

beta carbolines

RANUNCULUS TRICHOPHYLLUS CHAIX

Phylum: Magnoliophyta
Class: Magnoliopsida
Order: Ranunculales
Family: Ranunculaceae
Common name: Threadleaf crowfoot, water buttercup
Distribution: United States, United Kingdom
Habitat: Shallow calm waters, ditches, rivers, lakes, ponds
Description: This species has only submersed thread leaves that resemble buttercups. The flowers are white and 6–12 mm in diameter.
Compounds and activities: The ethanolic extracts of this plant species possess flavones and ent-labdanes, which have antiprotozoal and antipyretic activities. The extract displays appreciable trypanocidal activity against *Trypanosoma brucei rhodesiense*. It also possesses leishmanicidal potential (Orhan et al., 2006).

POLYGONUM SENEGALENSE MEIS W (= *PERSICARIA SENEGALENSIS*)

Phylum: Magnoliophyta
Class: Magnoliopsida
Order: Caryophyllales
Family: Polygonaceae

Common name: Knotweed

Distribution: Botswana, Malawi, Mozambique, Zambia, Zimbabwe, Egypt, Madagascar, South Africa

Habitat: Semiaquatic habitats, banks of dams, lakes, and rivers

Description: This is a nonwoody perennial, glabrous plant with green leaves that are simple and with entire margins.

Compounds and activities: From the leaf extracts of this species, 13 nonpolar flavonoid derivatives (chalcones, dihydrochalcones, and flavones) have been isolated. The plant extracts contain 2′-glucosyl-6′-hydroxyl-4′-methoxychalcone. From the internal aerial tissues, common flavonoids, quercetin, kaempferol, luteolin, and their glycosides have been isolated. The phenolic derivative, polygonomoisoflavonone, viz., 2-propen-1-one,1-(2,4-dihydroxy-3,6-dimethoxyphenyl-3-phenyl), isolated from this species has been found to possess antimalarial (antiprotozoal) activities, as it acts *in vitro* against *Plasmodium falciparum.* Two chalcones ($R_1 = R_2 = OCH_3$, RC = O, and $R_1 = H$, $R_2 = OH$, $R_3 = OCH_3$) isolated from the aerial exudates of this species were found to be active with IC_{50} values of 3.1 and 14.0 μm, respectively (Bero et al., 2009). The extracts of this species with sodium carbonate have been used in traditional medicines for treating rheumatic swellings, eruptive fever (antipyretic), and urethral discharge.

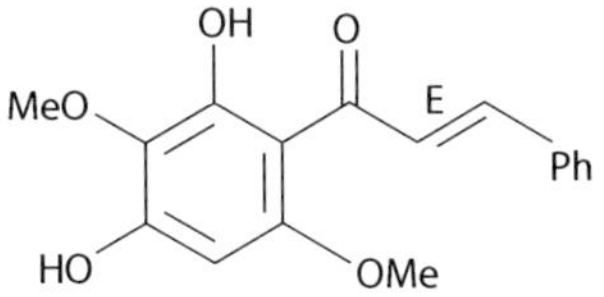

2-propen-1-one,1-(2,4-dihydroxy-3,6-dimethoxyphenyl-3-phenyl

POTAMOGETON PERFOLIATUS LINNAEUS

Phylum: Magnoliophyta

Class: Liliopsida

Order: Alismatales

Family: Potamogetonaceae

Common name: Clasping leaf pond weed, rock head grass
Distribution: Central and North America, Ireland, Europe, Asia, Australia
Habitat: Lakes, ditches, slow-flowing rivers, streams, brackish waters
Description: This is a common pond weed species, in which all leaves (8 cm long) are submersed, oval, and translucent. The inflorescence is with 6–15 flowers and fruits are olive-green.
Compounds and activities: The ethanolic extracts of this plant species possess flavones and ent-labdanes, which have antiprotozoal and antipyretic activities. The extract displays appreciable trypanocidal activity against *Trypanosoma brucei rhodesiense.* It also possesses leishmanicidal potential (Orhan et al., 2006).

PHRAGMITES AUSTRALIS (CAV.) TRIN. EX. STEUD

Phylum: Magnoliophyta
Class: Liliopsida
Order: Cyperales
Family: Poaceae
Common name: Common reed
Distribution: Cosmopolitan in distribution
Habitat: Wetlands, ponds, shallow waters
Description: This is a perennial subshrub or shrub growing to 4 m. The stem is erect with green foliage and white flowers, which are hermaphroditic. The rhizostomatous growth is seen in this species.
Compounds and activities: From the extracts of this species, seven flavonoid glycosides, one flavonone, seven simple phenolic compounds, and two simple nitrogen heterocyclic compounds have been isolated. All these compounds have shown antimalarial activities through their toxic and repellent activity of larvae and adults of mosquito *Culex pipiens*, thereby controlling the diseases caused by mosquito-borne pathogens. The ethanol extracts of leaves of this species were found to be most effective (LC_{50} 98.72), followed by its stems (LC_{50} 619.52). The petroleum extracts of this species, however, showed LC_{50} values of 60.06 (leaves) and 16.13 (stem). Further, the petroleum ether extract of the leaves showed 100% repellency at a dose

of 6.6 mg/cm^2 (Bream et al., 2009), and the above compounds also have shown antitumor properties. The decoction made from the leaves of this species cures bronchitis, cholera, and food poisoning. The stem is used as an antidote and antipyretic agent. The root extracts possess antiasthmatic, antipyretic, and sedative properties. The plant extract is also used internally for treating diarrhea, fever, vomiting, coughs, lung abscesses, and urinary tract infections. On the other hand, the extracts with gypsum are used externally for halitosis and toothache.

CYPERUS ARTICULATUS LINNAEUS

Phylum: Magnoliophyta Class: Liliopsida
Order: Poales Family: Cyperaceae
Common name: Jointed flatsedge
Distribution: Southern United States, Africa, Central Asia, Australia, South America, Africa
Habitat: Damp, marshy, and flooded areas along rivers and streams
Description: This plant (2 m long) grows from dividing rhizomes (2 × 1.5 cm). It has tall, green, and fibrous stems. There are small white flowers at the top of its stem.
Compounds and activities: This species is known to produce various compounds, such as alkaloids (flavonoids), polyphenols, saponins, tannins, and terpenes. Two sesquiterpenes, viz., corymbolone and mustakone, isolated from the chloroform extract of the rhizomes exhibited antiplasmodial properties (IC_{50} = 4.53 and 0.64 μm against NF 54 and IC_{50} = 8.14 and 1.15 μm against EN 36, respectively) (Bero et al., 2009). The cyperotundone and α-cyperone isolated from this species are known to be potential pain relievers, like aspirin and ibuprofen, besides possessing antimalarial properties. While the methanolic extract of this species possesses anticonvulsant properties, the decoction of its rhizomes has sedative activities. Traditionally, the extract of this species is used as a cough remedy, in addition to flavoring food and manufacturing cosmetics.

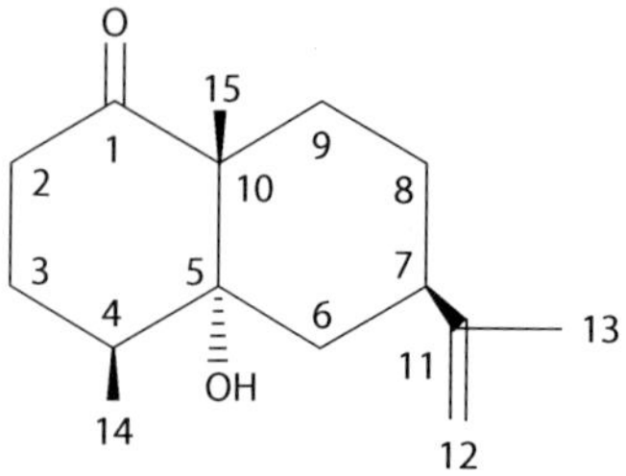

corymbolone (sesquiterpenes)

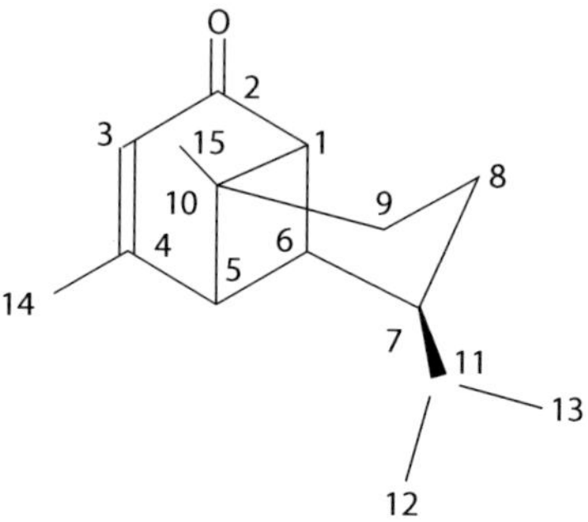

mustakone

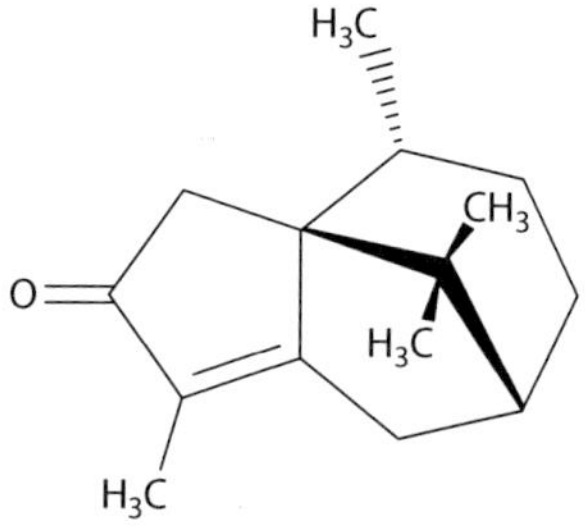

cyperotundone

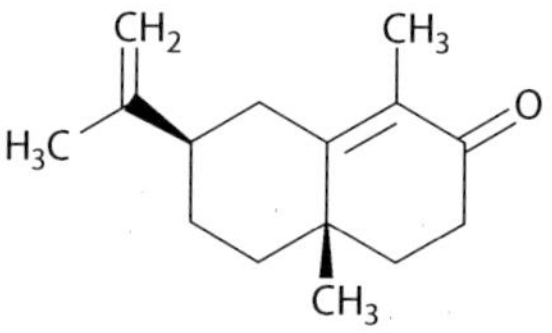

alpha cyperone

IRIS PSEUDACORUS LINNAEUS

Phylum: Magnoliophyta
Class: Liliopsida
Order: Liliales
Family: Iridaceae
Common name: Yellow flag, yellow iris
Distribution: Europe, West Asia, North Africa, Mediterranean
Habitat: Shallow river, edges of water courses, lakes, ponds

Description: Leaves (90 cm long) of this species are bluish green and linear, narrow, and sword shaped. Flowers are bright yellow or cream colored and are 7–10 cm wide.

Compounds and activities: The ethanol extract of leaves of this species has been reported to possess activities against the larvae of mosquito spp., viz., *Culex pipiens* and *Aedes caspius*, which are vectors of Japanese encephalitis and malaria, respectively. The 24 h LC_{50} values of these species are 10.36 and 16.43 mg/l, respectively. The isoflavonoids of this species possess antimicrobial activities inhibiting the growth of pathogens (Small and Catling, 1999).

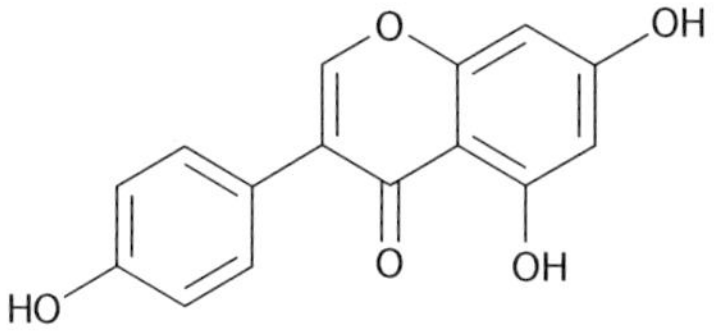

isoflavonoids

PISTIA STRATIOTES LINNAEUS

Phylum: Magnoliophyta Class: Liliopsida
Order: Alismatales Family: Araceae
Common name: Water cabbage, water lettuce
Distribution: All tropical and subtropical countries
Habitat: Freshwater ways, rivers, lakes, ponds

Description: This perennial monocot plant has thick and soft leaves that form a rosette. While the leaves are floating on the surface of the water, the roots are hanging and submerged below the floating leaves. The leaves (15 cm long) are light green with wavy margins. The flowers are dioceous. This plant may also reproduce by asexual methods.

Compounds and activities: The extracts of this plant species have been reported to possess antimicrobial, antimalarial, antihelmintic, antidermatophytic, antidiabetic, antipyretic, analgesic, and anti-inflammatory activities. The bioactive compounds isolated from the roots of this species include alkanes, flavonoids, sterols, stigmasta-4, 22-dien-3-one, stigmasterol, stigmasteryl stearate, and palmitic acid. Larvicidal activities against the larvae of *Culex* using crude (1) and chloroform (2) extracts of this species showed LC_{50} values of 159.50 μg/ml for 1 and 0.909 μg/ml for 2. The values of percentages of mortality at 40 μg/ml of extract were 16.67% (1) and 90.0% (2) (Mukhtar et al., 2004). This plant species also has several therapeutic functions. While the entire plant is antiseptic, antitubercular, and antidysenteric, its root extracts are emollient, laxative, and diuretic. Its leaf extracts also possess demulcent, refrigerant, diuretic, and antihyperglycemic functions. In traditional medicines, this plant has been used as a therapeutic agent in the treatment of flu and ulcerative conditions of the mouth and tongue.

4 Antihelmintic Compounds

INTRODUCTION

Helmintic infections may cause diseases such as anemia, malnutrition, eosinophilia, and pneumonia. Tannins and polyphenolic compounds produced by freshwater plants largely help animals by decreasing helminth infestation.

NYMPHOIDES HYDROPHYLLA (LOUR) O. KUNTZE

Phylum: Magnoliophyta
Class: Magnoliopsida
Order: Solanales
Family: Menyanthaceae
Common name: Crested floating heart
Distribution: Tropical Asia, Taiwan, Bangladesh
Habitat: Lakes, ponds, ditches
Description: This floating herb has a stem of 30 cm length. The rooting is from nodes. Leaves are heart shaped. Flowers are small, simple, and with five star-shaped or rounded petals.
Compounds and activities: This plant's seed powder possesses antihelmintic properties and may be consumed with honey. The edible portions of this plant, viz., stem, leaves, and fruits, have been reported to be useful for curing jaundice and fevers. The extracts of its stalks and leaves with oil are used for ulcers and insect bites (Swapna et al., 2011).

POLYGONUM GLABRUM (WILLD)

Phylum: Magnoliophyta
Class: Magnoliopsida
Order: Polygonales
Family: Polygonaceae
Common name: Common marsh, buckwheat
Distribution: India, Nepal, Sri Lanka, China, Africa, Afghanistan, Japan, Malaysia
Habitat: Semiaquatic, marshy areas
Description: The leaves are lanceolate, and their arrangement is alternate spiral. While the leaf base is acute, its margin is entire. The flowers are pink and are in terminal racemes.
Compounds and activities: The bioactive compound isolated from this species is 1,1-diethoxy propane. The extracts of leaves of this species have yielded phenolic compounds, tannins, and alkaloids. Compounds such as four new diesters of 2,3-dihydroxyisodrimeninol of this species have shown antihelmintic activities. The phenolic and alkaloid compounds of this species have shown antibacterial activities against *Proteus vulgaris* and *Bacillus subtilis.* The minimum inhibitory concentration (MIC) value of stem methanol extract and leaf ethanol extract of this species against *B. subtilis* was 1 mg/ml. Further, the MIC values of flower ethyl acetate extract and leaf methanol extract of this species against *Proteus vulgaris* were found to be 1 and 0.5 mg/ml, respectively (Jani et al., 2012). In another investigation, its 1,1-diethoxy propane content has been reported to possess antimicrobial activities to the extent of 65%.

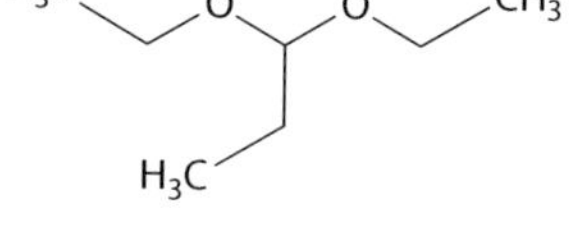

1,1-diethoxy propane

CAREX BACCANS NEAS

Phylum: Magnoliophyta | Class: Liliopsida
Order: Cyperales (Poales) | Family: Cyperaceae
Common name: Crimson seeded sedge
Distribution: Temperate and tropical Asia
Habitat: Swamps, moist places
Description: This species, a climbing evergreen herb, is known for its bright showy seed heads. It possesses dark green leaves (90 × 1 cm). The flower spike carries greenish flowering seed clusters that grow to 30 cm long and 10 cm wide.
Compounds and activities: The crude ethanolic root extract of this species was found to show complete inactivation and flaccid paralysis, leading to the death of a cestode parasite of domestic fowl, viz., *Raillietina echinobothridia* (Challam et al., 2012). The endophytic actinomycetes (*Streptomyces hydrogerans*) isolated from the roots of this species have, however, shown inhibitory activities against pathogenic bacteria and fungi.

5 Antiviral Compounds

The antiviral compounds are responsible for producing protective or therapeutic effects to clear a detectable advantage of the virus-infected host. The use of aquatic plants traditionally as antiviral components is certainly broader than their applications in modern drugs. The antiviral compounds isolated from the aquatic plants include alkaloids, flavol, flavonone, isoflavonone, etc.

ANKISTRODESMUS CONVOLUTUS CORDA, 1838

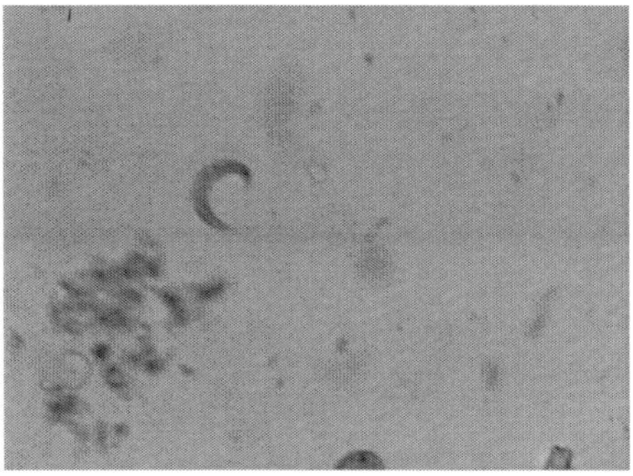

Phylum: Chlorophyta Class: Chlorophyceae
Order: Sphaeropleales Family: Selenastraceae
Distribution: Central Asia, Australia, New Zealand
Habitat: Ponds, lakes, rivers
Description: This species has solitary or groups of two to five cells (5 µm diameter, 25 µm long) that are sigmoid, twisted, or fusiform in shape. The apices of cells are sharply pointed, invariably twisted in opposite directions.
Compounds and activities: The methanol extracts of this species containing antiviral ACF 1 *a* and *b* have been found to be active against Epstein-Barr virus (EBV). This extract displayed low cytotoxicity and a potent effect in reducing cell-free EBV DNA (EC_{50} < 0.01 µg/ml) with a high therapeutic index (>28,000) (Kok et al., 2011).

CHARA VULGARIS LINNAEUS, 1753

Phylum: Chlorophyta
Class: Chlorophyceae
Order: Charales
Family: Characeae
Common name: Common stonewort
Distribution: South America, North America, Southwest Asia, Central Asia, Australia, New Zealand
Habitat: Ponds, puddles, ditches, rivers, lakes, brackish waters
Description: The annual or perennial species grows to a maximum height of 50 cm, and its axis measures 0.5–1.0 mm diameter. The plant is gray to dark green in color with seven to nine branches. Each of these branches has six to eight segments. The flowers are polymorphic and monoecious.
Compounds and activities: The bioactive sterol compounds isolated from this species are 24-ethyl cholest-5-en-3-ol, clionasterol (an isomer of sitosterol), 28-isofucosterol, and an isomer of sitosterol. The 24-ethyl cholest-5-en-3-ol possesses antiviral activities and prevents common cold and flu, HIV, and AIDs. Further, the extracts of this species have been reported to mitigate heart problems, high cholesterol, rheumatoid arthritis, tuberculosis, psoriasis, cervical cancer, asthma, bronchitis, chronic fatigue syndrome, and hair loss, besides boosting the immune system in humans.

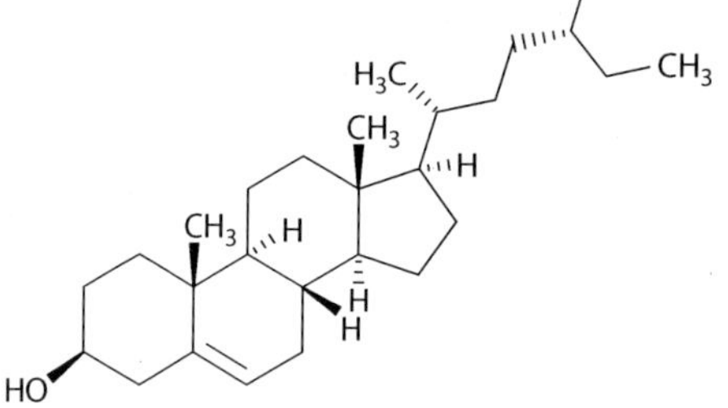

clionasterol

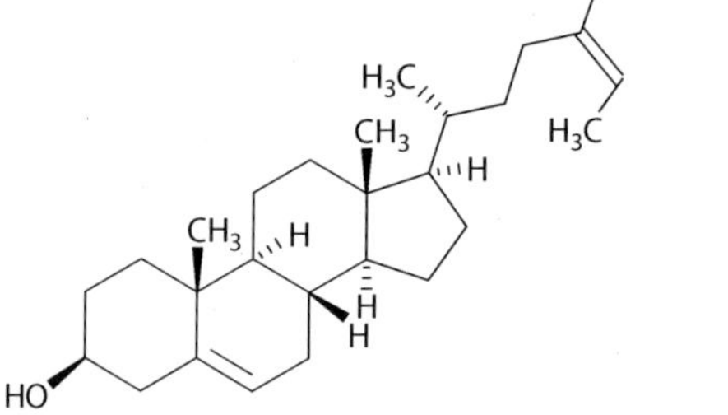

28-isofucosterol

SYNECHOCOCCUS ELONGATUS (NAGELI) NAGELI, 1849

Phylum: Cyanobacteria
Class: Cyanophyceae
Order: Synechococcales
Family: Synechococcaceae
Distribution: Europe, Southwest Asia, Central Asia
Habitat: Dirty ponds, eutrophic habitats
Description: The cells of this unicellular species are rod shaped and are without flagella or cilia. They are photoautotrophic and can swim or glide.
Compounds and activities: The methanol extracts of this plant species have been reported to show potent effects in reducing cell-free Epstein-Barr virus (EBV) DNA with a high therapeutic index. A fraction of the extract obtained from column chromatography was found to reduce the cell-free EBV DNA most effectively (EC_{50} = 2.9 μg/ml) with a therapeutic index of >69. On the other hand, with further fractionation by high-performance liquid chromatography (HPLC), its subfraction was found to be most effective in reducing the cell-free EBV DNA (EC_{50} = 1.38 μg/ml) with a therapeutic index of >14.5 (Kok et al., 2011).

ANABAENA SPHAERICA BORNET & FLAHAULT, 1888

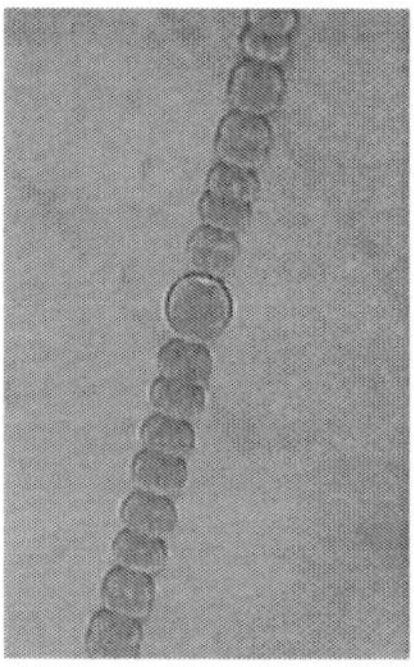

Phylum: Cyanobacteria
Class: Cyanophyceae
Order: Nostocales
Family: Nostocaceae
Distribution: Europe, Africa
Habitat: Freshwater habitats, rice fields
Description: This is a filamentous, nitrogen-fixing planktonic species. The filaments are unbranched, and the cells are arranged end to end. Heterocysts are present among cells of the filaments.
Compounds and activities: The bioactive compounds isolated from this species include carotenoids, phycocyanin, and several saturated and unsaturated fatty acids. The unsaturated fatty acids include oleic, linoleic, and linolenic acids. The unsaponifiable compounds isolated from this species are cholesterol, campesterol, and stigmasterol. The phycocyanin of this

species possesses antiviral, anticancer, and hypocholesterolemic activities besides serving as a potent inhibitor of human leukemia K562 cells and herpes simplex virus type 2 (HSV-2). Further, its extract has also been reported to possess antibacterial properties. The values of inhibition zone diameter (mm) for bacterial species, *S. epidermidis*, *S. aureus*, *S. haemolyticus*, *P. vulgaris*, and *S. sonnei*, were found to be 16, 17, 14, 15, and 16, respectively. The values for fungal species, *C. krusei*, *C. albicans*, *A. niger*, *A. fumigatus*, and *C. neoformans*, were 8, 7, 7, 6, and 9 mm, respectively (Ramamurthy et al., 2012).

NOSTOC ELLIPSOSPORUM RABENHORST EX BORNET & FLAHAULT, 1888

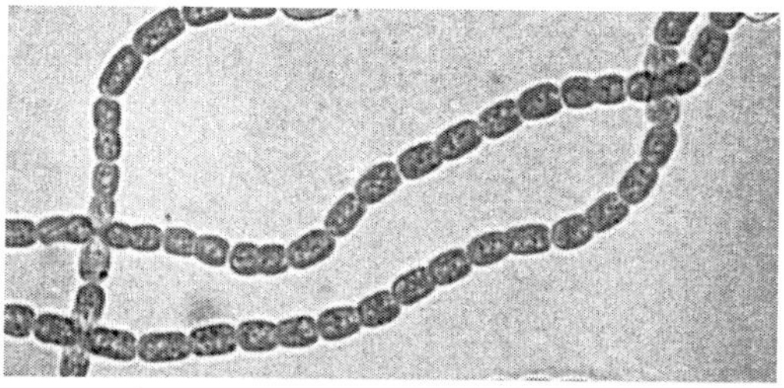

Phylum: Cyanobacteria
Class: Cyanophyceae
Order: Nostocales
Family: Nostocaceae
Distribution: Europe
Habitat: Freshwater habitats
Description: This species forms spherical colonies in which the filaments are embedded in a gelatinous substance.
Compounds and activities: The bioactive compound cyanoviridin-N (101 amino acid long, 11 kDa protein) isolated from this species has been reported to be virucidal and inactivate the strains of human immunodeficiency virus type 1 (HIV-1) (by interacting with mannose groups of envelope glycoproteins, gp 120, and blocking their interaction with target cell receptors), HIV-2, HSV-6, measles virus, simian immunodeficiency virus (SIV), and feline immunodeficiency virus (FIV) (Singh et al., 2005).

SCYTONEMA VARIUM KUTZING, 1849

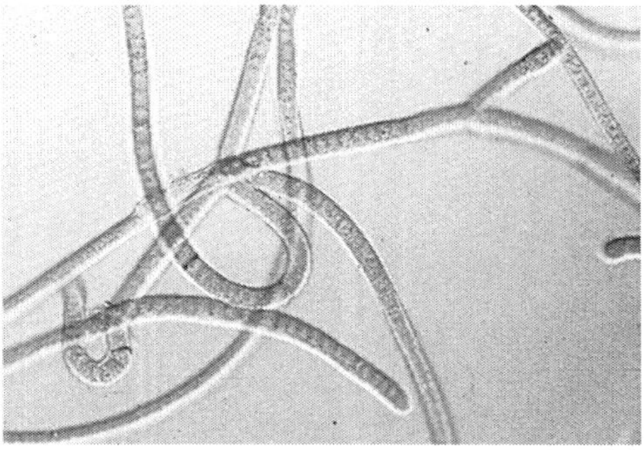

Phylum: Cyanobacteria
Class: Cyanophyceae
Order: Nostocales
Family: Scytonemataceae
Distribution: Pacific Islands
Habitat: Freshwater and brackish water habitats
Description: The filaments of this species may be branched and form mats. The trichomes are isopolar, cylindrical, and uniseriate. The heterocysts are solitary.
Compounds and activities: The bioactive compounds, viz., scytovirin and cyanovirin N, isolated from this species have shown inhibitory effects on HIV protein (Bokesh et al., 2003). Scytovirin has also displayed potent anticytopathic activity against laboratory strains and primary isolates of HIV-1 by interacting with oligosaccharides containing α1-2 and α1-6 tetramannose units of envelope glycoproteins, gp 120, gp 160, and gp 41.

LYNGBYA LAGERHEIMEII (GOM.) ANAGN. ET. KOM

Phylum: Cyanobacteria
Class: Cyanophyceae
Order: Oscillatoriales
Family: Oscillatoriaceae
Distribution: Europe
Habitat: Shallow waters
Description: The filaments of this species are without heterocysts.
Compounds and activities: The crude extract of this species has yielded HIV inhibitors that protect human lymphoblastoid T cells from the cytopathic effects of HIV infection (Singh et al., 2005).

SPIRULINA MAXIMA SP. NOV. (= *ANTHOSPIRA MAXIMA*)

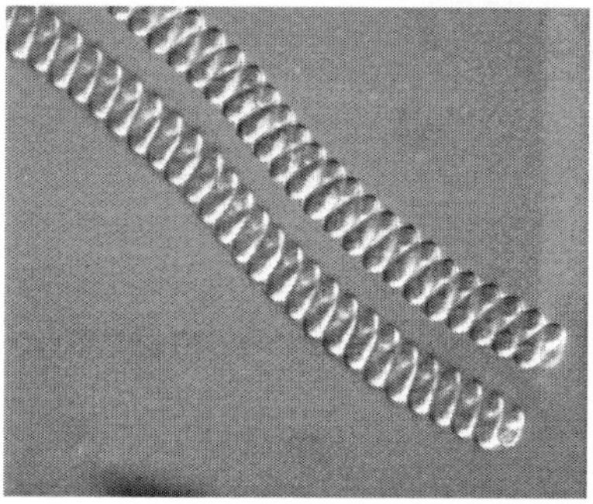

Phylum: Cyanobacteria
Class: Cyanophyceae
Order: Pseudanabaenales
Family: Pseudanabaenaceae
Distribution: Central America, Mexico, Africa, Central Asia, North and South America
Habitat: Alkaline freshwater habitats, lakes
Description: This photosynthetic and multicellular species has spiral-shaped filaments.
Compounds and activities: The hot-water extract of this species was found to inhibit the infection of HSV-2, pseudorabies virus (PRV), human cytomegalovirus (HCMV), and HSV-1, and the 50% effective inhibition doses (ED_{50}) were 0.069, 0.0103, 0.142, and 0.333 mg/ml for each virus, respectively. The highest antiviral activity (for HSV-2 with a selective index of 128) was with methanol-water (3:1), which was due to its highly polar compounds (Corona, 2002). The extracts of this species have also shown antimicrobial, antihyperlipidemic (decreased postprandial lipemia), antihyperglycemic, antihypertensive, and hepatoprotective activities. The presence of about 70 powerful phytonutrients of this species make it a good food supplement, and these compounds have been reported to reduce systolic and diastolic blood pressure in humans.

RANUNCULUS SIEBOLDII MIQ AND *RANUNCULUS SCELERATUS* LINNAEUS

Phylum: Magnoliophyta
Class: Magnoliopsida
Order: Ranunculales
Family: Ranunculaceae

Ranunculus sieboldii

Common name: Swamp buttercup

Distribution: China

Habitat: Wet habitats, rivers

Description: In this herbaceous dicot species, the leaves are palmate and the flowers, which are medium to large sized, are showy.

Ranunculus sceleratus

Common name: Cursed buttercup

Distribution: North America, Europe, Asia

Habitat: Ponds, stream banks, wet and moist habitats

Description: This annual herb grows to 0.5 m and has leaves with small blades that are deeply lobed or divided into three leaflets. The flowers, which measure a few millimeters long, have three to five yellow petals.

Compounds and activities: The antiviral compounds isolated from these species include apigenin 4′-O-α-rhamnopyranoside, apigenin 7-O-β-glucopyranosyl-4-O-α-rhamnopyranoside, tricin-7-O-β-glucopyranoside, tricin, and isoscopoletin. These compounds have been reported to show inhibitory activities against hepatitis B virus (HBV) replication. The compound isoscopoletin showed significant inhibitory effects on the expression of HBs Ag and HBe Ag secretion from HepG2.2.15 cells. The compound protoatechuyl aldehyde of this species has acted against herpes simplex virus type 1 (HSV-1). The above compounds have great potential in traditional application of HBV therapy (Li et al., 2005). These plants have also been found to possess anti-inflammatory, antihistaminic, and antimicrobial activities and are used in traditional medicines for the treatment of jaundice, hepatitis, asthma, rheumatic arthritis, stomachache, and toothache.

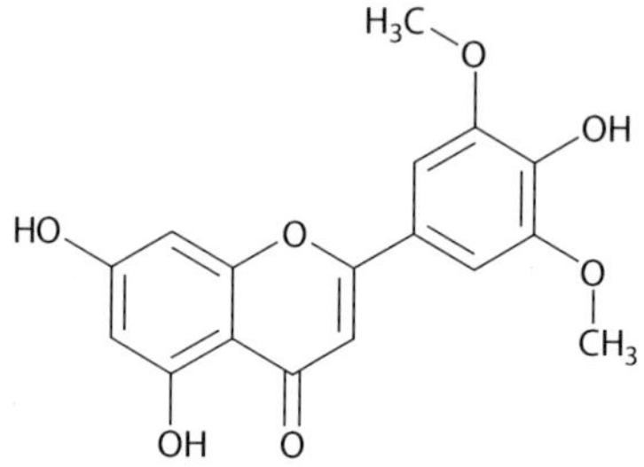

tricin

FALLOPIA JAPONICA HOUTT

Phylum: Magnoliophyta
Class: Magnoliopsida
Order: Caryophyllales
Family: Polygonaceae
Common name: Japanese knotweed
Distribution: East Asia, North America, Europe
Habitat: Ponds
Description: This large herbaceous perennial plant grows to a maximum height of 4 m. Its stems are hollow and leaves (15 × 12 cm) broad and oval with a truncated base and entire margin. Flowers are small, white, and in erect racemes (15 cm long).
Compounds and activities: The 70% ethanol extracts of this species contain the most active compounds, viz., (E)-resveratrol, 5,7-dimethoxyphthalide, (+)-catechin, and emodin-8-O-β-D-glucopyranoside, showing fairly strong antiviral activity against HIV-1-induced cytopathic effects in C8166 lymphocytes at noncytotoxic concentrations with EC_{50} values of 4.37 ± 1.96, 19.97 ± 5.09, 14.4 ± 1.34, and 11.29 ± 6.26 μg/ml and therapeutic index values of 8.12, >10.02, >13.89, and >17.71, respectively. The bioactive anthroquinones isolated from the roots of this species, viz., physcion, emodin, citreorosein, anthroglycoside B, and resveratrol, possess antityrosinase potency and are of great use as skin-whitening agents. Compounds physcion and emodin have also been reported to possess antidiabetic (type 1 diabetes) and anti-inflammatory activities.

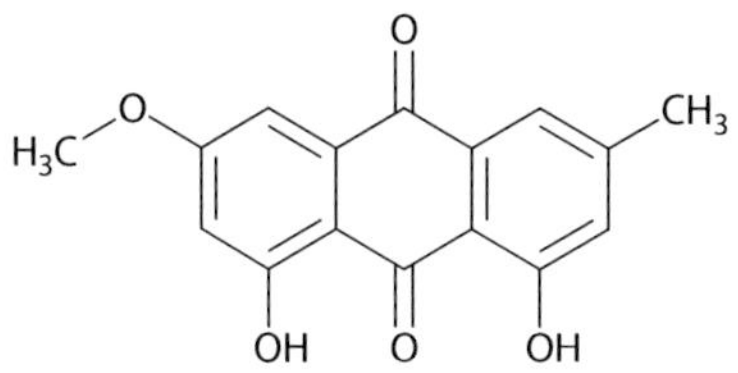

physcion

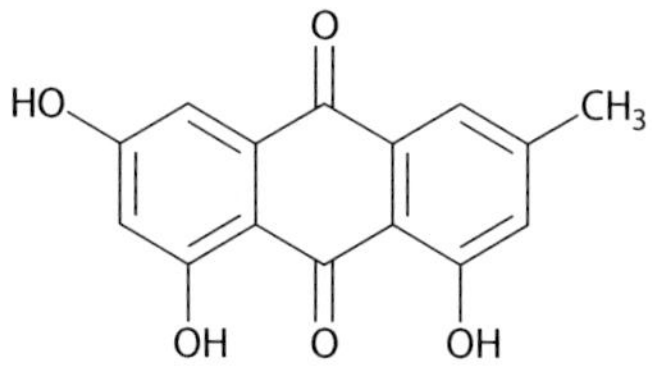

emodin

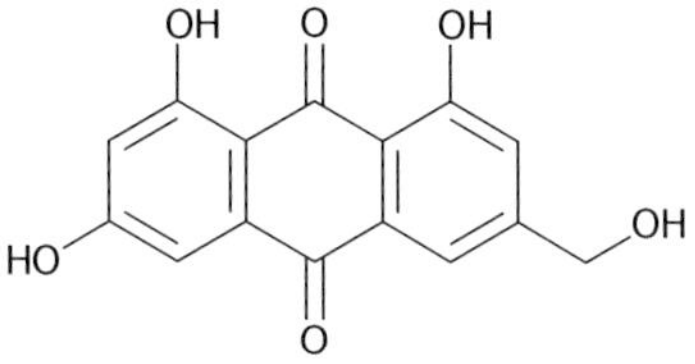

citreorosein

NEPTUNIA OLERACEA LOUR

Phylum: Magnoliophyta
Class: Magnoliopsida
Order: Fabales
Family: Fabaceae
Common name: Water mimosa
Distribution: Mexico to northern South America
Habitat: Still waters
Description: This perennial, nitrogen-fixing leguminous species has a stem (15 cm long) with aerenchyma that makes it float in water. Though the stem is short, it spreads in water to an even 1.5 m long. The sensitive leaves close up when they are touched. The leaf segments have 10–14 small oblong leaflets arranged in opposite pairs. The small greenish yellow flowers are found crowded into feathery or bicular inflorescences.
Compounds and activities: The leaf extracts of this species possess an active bioactive compound, viz., pheophorbide A, which has antitumor activities and inhibits the activation of tumor promoter-induced Epstein-Barr

virus (EBV) (Nakamura et al., 1996). Its flavonoids, triterpenoids, and tannins have shown hepatoprotective activity in carbon tetrachloride-induced rats. Further, this species has also exhibited the highest antioxidant activity (13.1 mg BHA-butylhydroxyanisol-equivalent/g dry weight). The juice of its stem and roots is used in traditional medicines for treating yellow fever and Guinea worm infection.

ALISMA ORIENTALIS (SAM) JUZEP

Phylum: Magnoliophyta Class: Liliopsida
Order: Alismatales Family: Alismataceae
Common name: Oriental water plantain
Distribution: Europe, Middle East, Egypt, China
Habitat: Damp ground and shallow water
Description: It is a perennial floating or submerged plant that grows to a maximum height of 120 cm. The leaf stalk is unbranched with a single leaf that has a prominent vest. The three petaled hermaphrodite flowers are white, pink, or green and are in clusters in whorls of 3 to 10. The plants regrow by their bulbous bases.
Compounds and activities: The bioactive compounds isolated from this species possess antidiabetic (antihyperglycemic), antihepatitive, and antidiuretic properties. The rhizomes of this species contain triterpenoid alisol O, alisol A-24 acetate, 25-anhydroalisol A, 13β, 17β-epoxyalisol A, alisol B-23-acetate, alisol, and alisol F 24 acetate. Except alisol O, all other compounds showed inhibitory activity *in vitro* on hepatitis B virus (HBV) surface antigen (HBs Ag) secretion of the Hep G2.2.15 cell line with IC_{50} values of 2.3, 11.0, 15.4, 14.3, 0.6, and 7.7 μm, respectively, and on HBVe antigen (HBe Ag) secretion with IC_{50} values of 498.1, 17.6, 41.0, 19.9, 8.5, and 5.1 μm, respectively (Jiang et al., 2006). The triterpenes, alisol M 23-acetate and alisol A 23-acetate, of this species have shown antihyperglycemic

effects. This species can also downregulate bikunin mRNA expression, decrease the calcium oxalate formation in rat kidney, and inhibit renal stone formation in rat urolithiasis model (Cao et al., 2004). Dried rhizome of this species is used in traditional medicines as a crude drug for diuretics, diabetes, and hypolipidemic disorders.

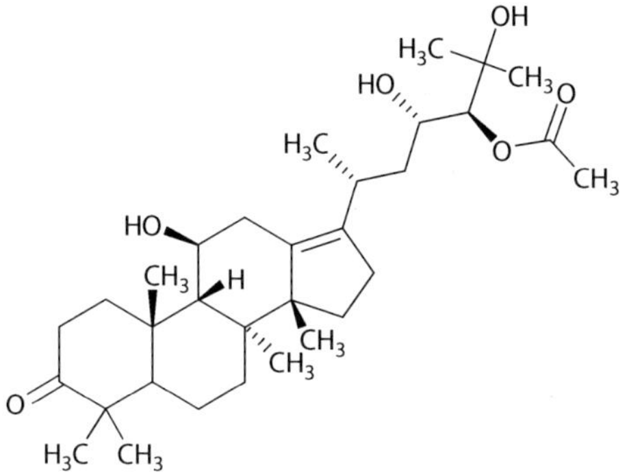

alisol A-24 acetate

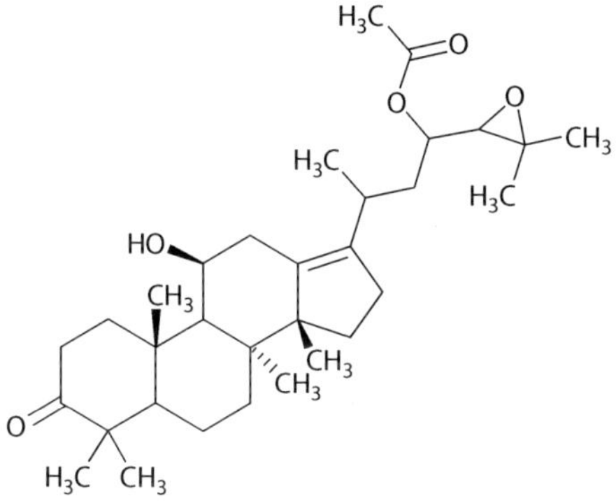

alisol B-23 acetate

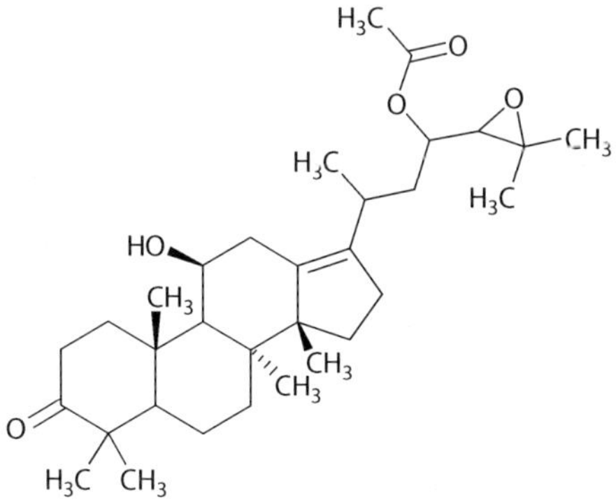

alisol

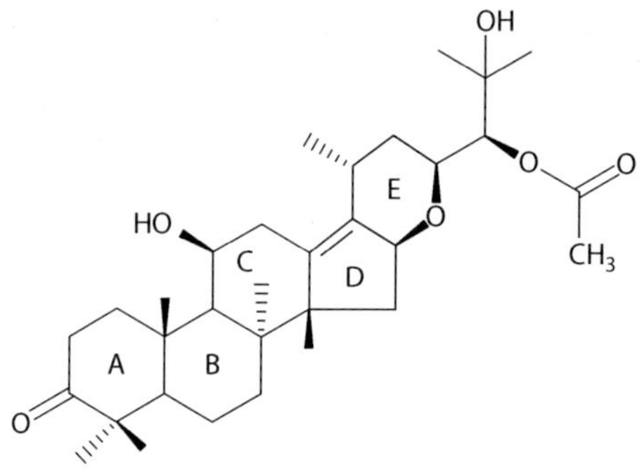

alisol F24 acetate

POTAMOGETON FERRUGINEUS POIR

Phylum: Magnoliophyta
Class: Liliopsida
Order: Alismatales
Family: Potamogetanaceae
Common name: Long-leaved pond weed, London pond weed
Distribution: North America, Europe, Asia, Americas
Habitat: Ponds, lakes, ditches, streams, eutrophic rivers, flowing waters
Description: This perennial, submersed macrophyte grows to 2 m and has branched rhizomes. The leaves (15 × 4 cm), which are alternate, entire, narrow, and lance-like, are submerged and floating. The green-brown flowers, which form small inflorescences in dense cylindrical spikes, are floating above water.
Compounds and activities: The extracts of this species possess antiviral properties.

POTAMOGETON MALAIANUS MIQ

Phylum: Magnoliophyta Class: Liliopsida
Order: Alismatales Family: Potamogetanaceae
Common name: Pond weed
Distribution: China, Central and South America, Central Asia
Habitat: Aquatic habitats, ponds
Description: In this submersed species, the leaves (7 × 2 cm) are memberanous, light green, and linearly elliptical.
Compounds and activities: The methanol extract of this species has yielded bioactive compounds such as new furanoid labdone diterpenes, viz., potamogetonyde and potamogetonol, and two known compounds, viz., potamogetonin and 15,16-epoxy-12-oxo-8,13,14-labdatrien-20,19-olide. Among them, the compounds shown below exhibited cytotoxity against the vero cell line with respective IC_{50} values of 31 and 28 μg/ml. These compounds showed potent antiviral (HSV-1) activity with IC_{50} values of 8 and 3 μg/ml respectively. However, all compounds showed antimalarial activity and cytotoxicity toward mosquito larvae with IC_{50} values ranging from 11 to 72 μg/ml (Kittakoop et al., 2001).

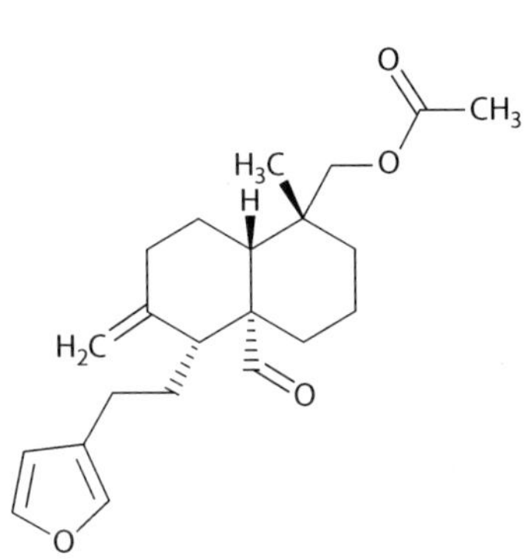

potamogetonyde

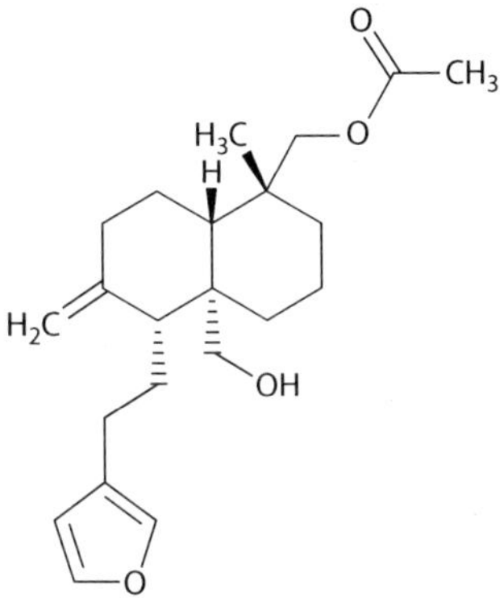

potamogetonol

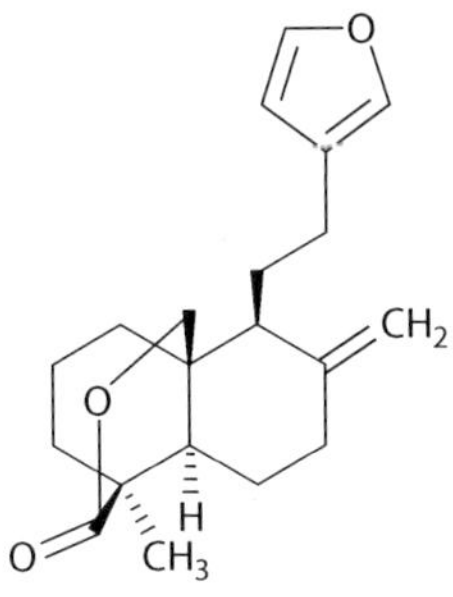

potamogetonin

6 Anticancer Compounds

Cancer in humans has been reported to occur due to errors in DNA and RNA replications. Anticancer drugs have largely been obtained from terrestrial plants, aquatic blue-green algae, and filamentous fungi. Research on the investigation of anticancer compounds from aquatic macrophytes is gaining momentum nowadays. The bioactive compounds primarily responsible for cytotoxic and proliferative activities include vinca alkaloids, epipodophyllotoxins, taxanes, and camptothecins. Other similar compounds identified are terpenoids, steroids, lignans, quinones, flavans, alkaloids, amides, and peptides, which, isolated from certain species of aquatic plants, have been reported to possess cytotoxic, antiproliferative, and anticancer activities for treating tumors, lymphomas, leukemias, etc.

HYDRODICTYON RETICULATUM (LINNAEUS) LAGERHEIM, 1883

Phylum: Chlorophyta
Class: Chlorophyceae
Order: Sphaeropleales
Family: Hydrodictyaceae
Common name: Water net
Distribution: Europe
Habitat: Ponds, lakes, rivers
Description: The cells of this species are cylindrical to oblong-oval. Uniquely, three to four cells are found connected at edges to form six-sided meshes that are clearly visible in larger colonies.
Compounds and activities: The bioactive compounds isolated from this species are 24-epicastasterone (24 epi cs) and 28 homo cs (natural brassinosteroids), which are currently used for treating hyperproliferative breast and prostate cancers. Further, the hexanic extracts of this species have anti-inflammatory and smooth muscle relaxing properties, i.e., antispasmodic, anticholinergic, and antihistaminic effects (Gutierrez and Solis, 2007).

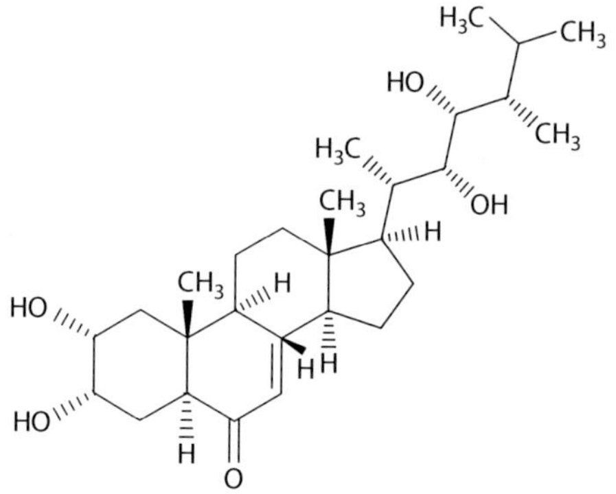

24-epicastasterone

CHLAMYDOMONAS REINHARDTII P.A. DANG

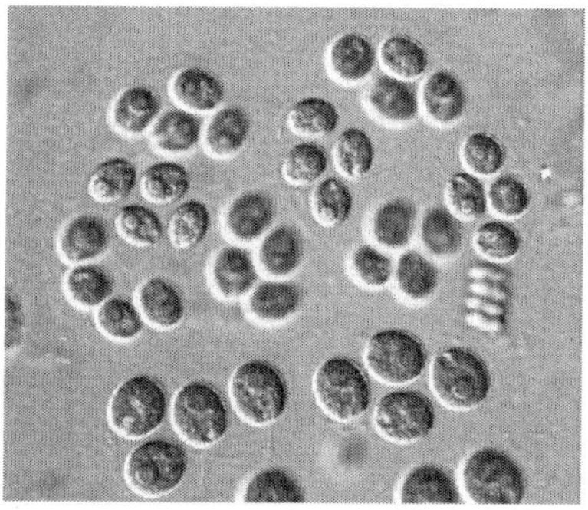

Phylum: Chlorophyta
Class: Chlorophyceae
Order: Volvocales
Family: Chlamydomonadaceae
Distribution: Throughout world
Habitat: Freshwater habitats
Description: This highly adaptable and single-celled species (10 μm diameter) swims with two flagella. A cup-shaped chloroplast, a large pyrenoid, and an eye spot are present in each cell.
Composition and activities: The B-type cyclin-dependent kinase produced by this species regulates the activity of tumor suppressor protein Rb and cell cycle progression. Further, the wall-less cw 15 mutant of this species, owing to its enzymes, namely, topoisomerases I and II, could advantageously replace mammalian cells in the screening of anticancer drugs (Maucourt et al., 2002). Other bioactive compounds of this species, viz., pinolenic acid and coniferonic acid (which are also present in pinus seed oils), have been reported to prevent osteoporosis. Its peptides have wide applications as additives in food industries and in the production of human therapeutic

proteins. The methanolic and hexanolic extracts of this plant have shown antibacterial activities against *S. aureus*, *Staphylococcus epidermidis*, *B. subtilis*, *E. coli*, and *Salmonella typhi*.

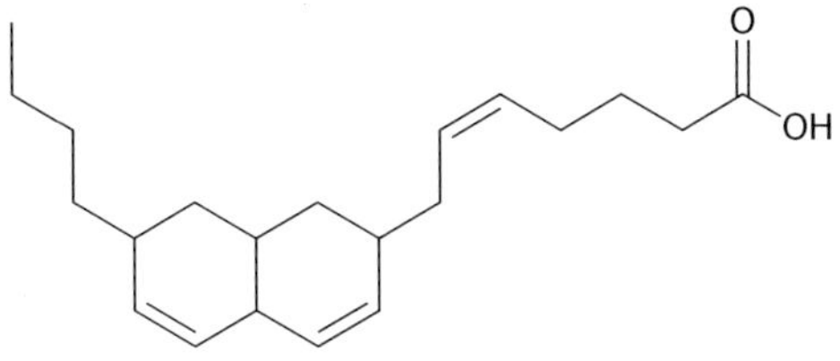

pinolenic acid

SPIRULINA FUSIFORMIS VORONIKHIN, 1934

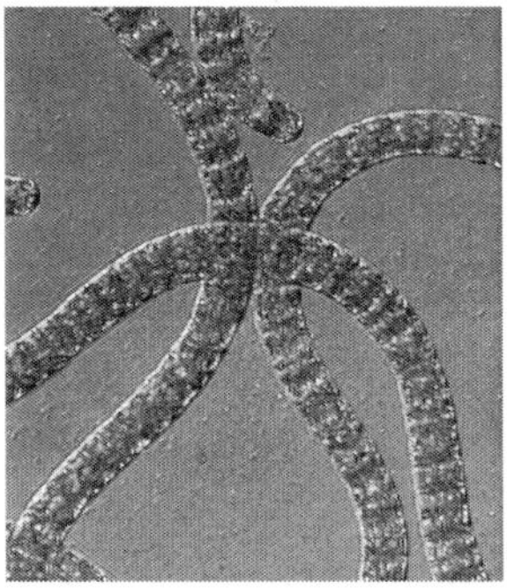

Phylum: Cyanobacteria Class: Cyanophyceae
Order: Pseudanabaenales Family: Pseudanabaenaceae
Distribution: Tropical and subtropical countries
Habitat: Freshwater habitats
Description: This free-floating, multicellular, filamentous species shows polymorphism; i.e., it changes its shape, color, and other characteristics according to environmental changes.
Compounds and activities: The extracts of this species show anticancer activities as oral supplementation inhibits chemically induced carcinogenesis. Its vitamin A content has shown chemopreventive activity in reversing oral cancer (leukoplakia) (Singh et al., 2005). The c-phycocyanin (blue-colored fluorescent protein) pigment of this species is a potent profibrinolytic protein in the vascular endothelial system. This species has also been reported to possess antiviral hepatoprotective, antioxidant, immunomodulatory, analgesic, antipyretic, and ulcerogenic properties (Rasool et al., 2008). The high content of K- and r-linolenic acid of this species has blood pressure-lowering effects.

LYNGBYA SP.

Phylum: Cyanobacteria
Class: Cyanophyceae
Order: Oscillatoriales
Family: Oscillatoriaceae
Habitat: Freshwater habitats

Description: This macroscopic filamentous, unidentified species has thick solitary filaments that are usually connected to thick layered leathery strata (mats). The trichomes (5 μm) are isopolar, straight, or fairly waved. The discoid cells are either barrel shaped or cylindrical.

Compounds and activities: This unidentified species has been reported to be a source of neurotoxins and drugs, and its bioactive compound, pahayokolide A, showed inhibitory activities against a number of cancer cell lines besides showing antimicrobial activities against *Bacillus* spp. and yeast, *Saccharomyces cerevisiae*.

OSCILLATORIA LIMNETICA LEMMERMANN, 1900

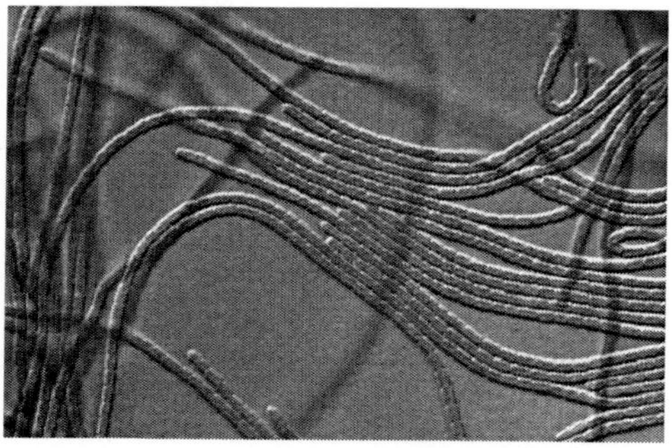

Phylum: Cyanobacteria
Class: Cyanophyceae
Order: Oscillatoriales
Family: Oscillatoriaceae
Distribution: Europe, Southwest Asia, Australia, New Zealand
Habitat: Freshwater habitats

Description: This species is motile by gliding movements. The trichome is straight and unbranched. The photosynthetic pigment chlorophyll *a* is present in the cells that are vegetative.

Compounds and activities: The bioactive compounds isolated from this species include carotenoids, phycocyanin, saturated and unsaturated fatty acids, and unsaponifiable compounds. The hexane extracts of this plant have also shown smooth muscle-relaxing properties. The unsaturated fatty acids isolated include oleic and linoleic acids. The unsaponifiable compounds of this species include cholesterol, campesterol, and stigmasterol. The phycocyanin (0.18 mg/ml) of this species inhibits the growth of human leukemia k562 cells besides possessing antiviral properties against herpes simplex virus type 2 and hypocholesterolemic activities. Further, the hexane extract of this species possesses smooth muscle relaxing activity, i.e., antispasmodic, anticholinergic, and antihistamine properties (Gutierrez and Solis, 2007).

LYNGBYA HIERONYMUSII LEMMERMANN

Phylum: Cyanobacteria | Class: Cyanophyceae
Order: Oscillatoriales | Family: Oscillatoriaceae
Distribution: Europe, Asia, Australia, New Zealand
Habitat: Estuaries and freshwaters
Description: Filaments are solitary and straight normally. They are sometimes slightly bent and twisted. Loose bundles may also be seen. Many small gas vacuoles are found scattered in the cell. Trichomes are brownish green to olive-green. Cells measure 14 μm (width) and 21 μm (length).
Compounds and activities: Extracts of this species have been reported to enhance antitumor activity from 20 to 45% with an increase in concentration (Ghazala et al., 2009).

PHORMIDIUM TENUE (MENEGHINI) GOMONT, 1892

Phylum: Cyanobacteria | Class: Cyanophyceae
Order: Oscillatoriales | Family: Phormidiaceae
Distribution: Ireland, Europe, Southwest Asia, Central Asia, Australia, New Zealand
Habitat: Freshwater habitats
Description: The colonies of this species, which are bright blue-green, form a leathery mat. The trichome is straight or slightly flexuous, and attenuated toward the apex. The cells measure 2–5 μm length and 1–2 μm width. The end cell is long and conical. A thin and gelatinous sheath is present.
Compounds and activities: Several diacylglycerols of this species have been reported to inhibit chemically induced tumors in test animals. Its phycocyanin has also been reported to inhibit cyclooxygenase-2 (COX-2), which is overexpressed in breast cancer cells (Singh et al., 2005). Its sulfonic acid-containing glycolipid protects human lymphoblastoid T cells from the cytopathic effect of HIV infection.

NOSTOC LINCKIA (ROTH) BORNET, 1880

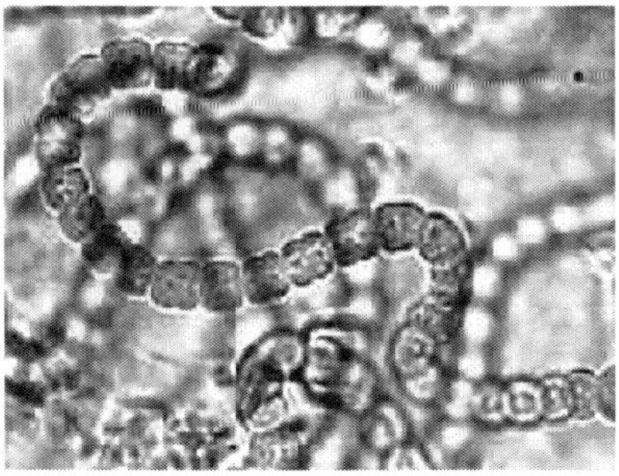

Phylum: Cyanobacteria | Class: Cyanophyceae
Order: Nostocales | Family: Nostocaceae

Distribution: Europe, Central Asia, Australia, New Zealand, Pacific Islands

Habitat: Freshwater habitats, brackish waters, and marine environments

Description: The filaments of this species are blue-green, brown, or violet, fluxuous or slightly coiled, and highly entangled. They form free-floating spherical colonies embedded in a gelatinous substance. The cells (3–4 μm) in the filaments are barrel shaped. The heterocysts (5–7 μm wide) are subspherical.

Compounds and activities: The bioactive compounds isolated from this species are 1,37-octatriacontadiene (octadecomol) and borophycin (boron-containing metabolite). The latter is cytotoxic, acting against human epidermoid carcinoma (LoVo) and colorectal adenocarcinoma (KB) cell lines. The fatty acids, viz., oleic and linoleic acids, isolated from this species possess antidiabetic properties (Boopathy and Kathiresan, 2010).

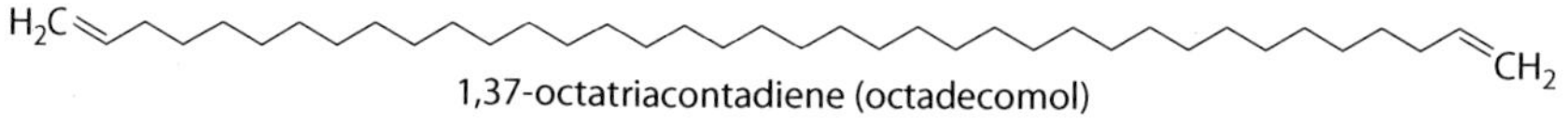

1,37-octatriacontadiene (octadecomol)

NOSTOC SP.

Phylum: Cyanobacteria
Class: Cyanophyceae
Order: Nostocales
Family: Nostocaceae
Habitat: Freshwater habitats

Description: This unidentified species forms colonies made of moniliform filaments. The cells are covered in a gelatinous sheath.

Compounds and activities: The bioactive compound isolated from this unidentified species is cryptophycin, which is cytotoxic acting against human tumor cell lines. This compound may be a promising drug ingredient in many cancer therapies (Trimurtulu et al., 1994).

ENHYDRA FLUCTUANS LOUR

Phylum: Magnoliophyta
Class: Magnoliopsida
Order: Asteriales
Family: Asteriaceae

Common name: Marsh herb, watercress

Distribution: India, Bangladesh, Sri Lanka, Southeast Asia, tropical Africa

Habitat: Semiaquatic habitats

Description: It is a prostrate, spreading annual herb floating in water. The stem is about 50 cm long, and rooting is at the nodes. Leaves are stalkless and linear-oblong. Flowers are white or greenish white.

Compounds and activities: A variety of bioactive compounds, such as β-carotene, saponins, myricil alcohol, kaurol, cholesterol, sitosterol, glucoside, germacranolide, enhydrin, fluctuanin, and fluctuandin, have been isolated. Its flavoids, viz., baicalein 7-O-glucoside and baicalein 7-O-diglucoside, have been reported to inhibit Ehrlich's ascites carcinoma (EAC) in mice (Sannigrahi et al., 2010). Its leaf extracts have also been reported to cure leukoderma, bronchitis, paralysis, epilepsy, and convulsion, besides possessing laxative, antibilious (torpidity of liver), demulcent, analgesic, and anti-inflammatory properties (Sannigrahi et al., 2011).

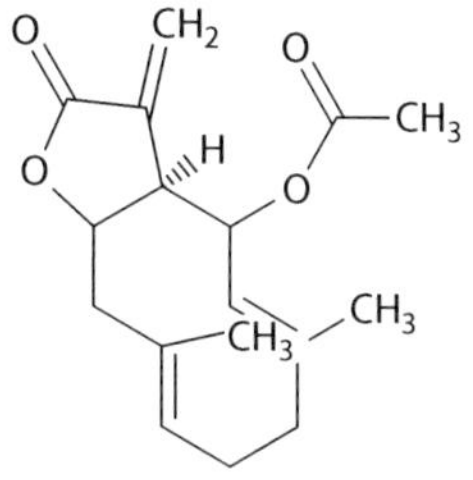

germacranolide

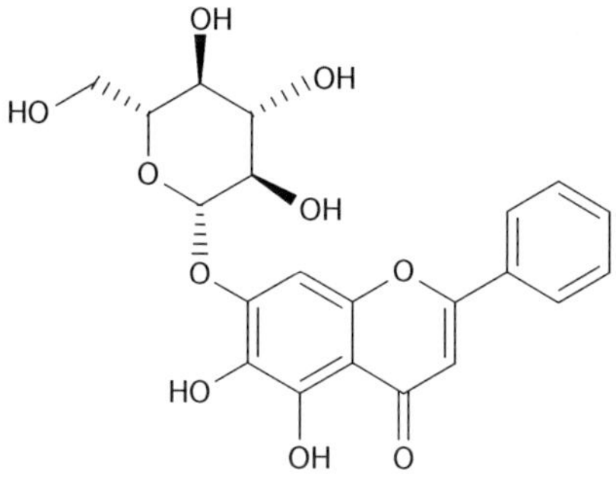

baicalein 7-O-glucoside

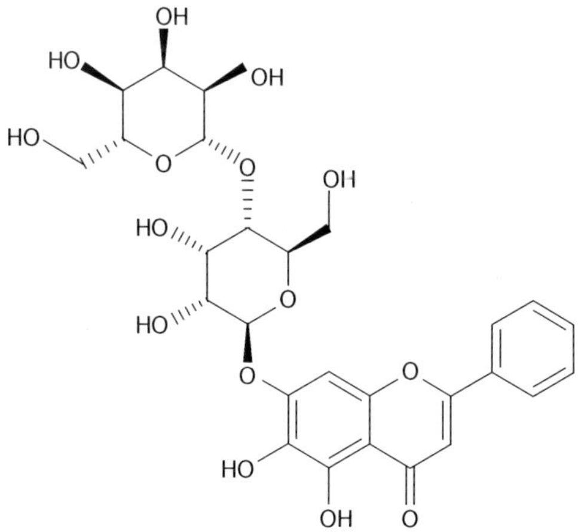

baicalein 7-O-diglucoside

ETHULIA CONYZOIDES LINNAEUS

Phylum: Magnoliophyta
Class: Magnoliopsida
Order: Asteriales
Family: Asteriaceae
Common name: Bungbungtit
Distribution: Ethiopia, Egypt, tropical Central Asia, tropical Africa
Habitat: Wet soil along rivers and swamps, riverbanks
Description: This erect annual or perennial herb grows to a maximum height of 120 cm. The stem is woody near the base and branched at the top. Leaves (5–10 cm long) are linear-elliptic to lanceolate with entire or serrated margins. The florets (5–6 mm diameter) are whitish to purple-blue.
Compounds and activities: From the leaves of this species, a monoterpene compound, viz., 5-methyl coumarin (acetophenone derivative), possessing moderate cytotoxic properties (40–50% activity at 500 μg/ml) has been isolated (Sowemimo et al., 2009). This species is therefore considered in traditional recipes for the treatment of cancer. In traditional medicines, the extract of this species is squeezed into the eyes for relief from headache. Further, the extract with red pepper cures constipation. Leaves in food are known to prevent abortion.

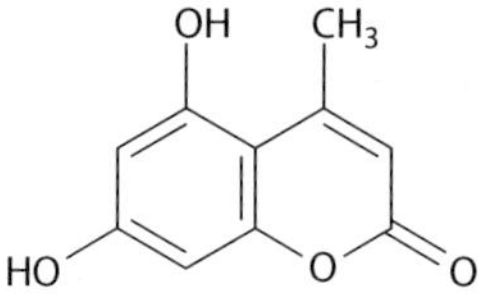

5-methyl coumarin

POLYGONUM LEPATHIFOLIUM LINNAEUS

Phylum: Magnoliophyta
Class: Magnoliopsida
Order: Caryophyllales
Family: Polygonaceae
Common name: Curly-top knotweed
Distribution: Europe, Asia, North America
Habitat: Wet to moist habitats
Description: This is an erect forb growing to a maximum height of 1.5 m and having unhairy stems with nodes. The leaves (30 cm long) are alternate and lanceolate. The four- or five-part flowers (3 mm long) are white to pinkish and are in dense clusters or inflorescences of 5 cm long.
Compounds and activities: This plant has yielded bioactive flavonoids, viz., 5-methoxy-6,7-methylenedioxy flavones, 5-hydroxy-7-methoxy flavonone (pinostrobin), and phenylpropanoid esters of sucrose, vanicoside B, and lapathoside A. Among them, compounds phenylpropanoid esters of sucrose, vanicoside B, and lapathoside have shown antitumor and cancer chemopreventive activities. The compound pinostrobin has shown antileukemic activity (Midori et al., 2001).

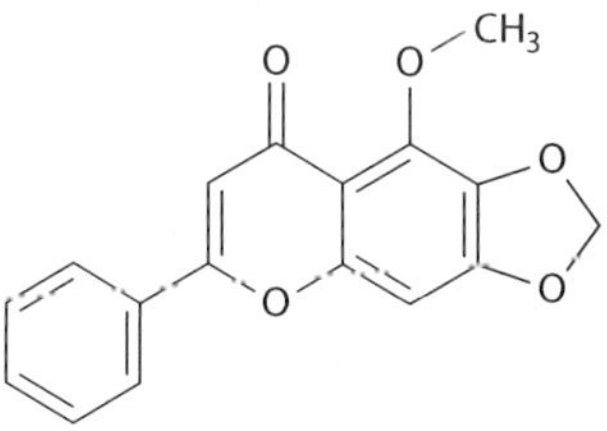

5-methoxy-6, 7-methylenedioxy flavones

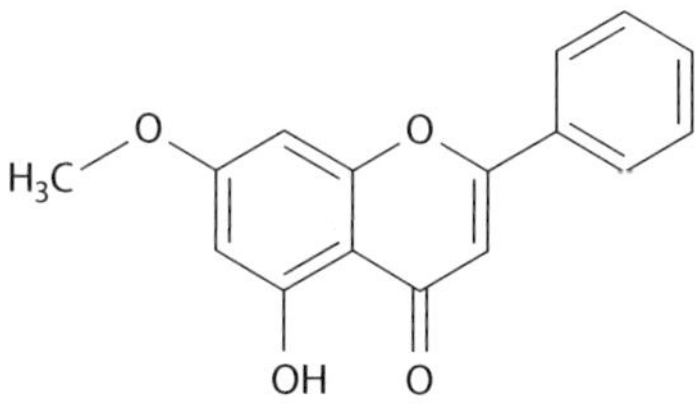

5-hydroxy-7-methoxy flavonone

POLYGONUM BARBATUM LINNAEUS

Phylum: Magnoliophyta
Class: Magnoliopsida
Order: Polygonales
Family: Polygonaceae
Common name: Joint weed
Distribution: Southeast Asia, tropical Australia, tropical Africa
Habitat: Freshwater or brackish water pools, marshy sites, tidal rice fields
Description: It is a perennial erect herb growing to a maximum height of 1 m. Roots are fibrous and white or brown. The stem is erect, round or hollow, glabrous, and thickened at the nodes. Leaves are alternate, spiral, stalked, and >2 cm long/wide. The margins of leaves are entire and their apices are acute. White or pink flowers are hermaphroditic and are grouped together into a terminal spike that is stalked.
Compounds and activities: The petroleum ether and methanol extracts of this species showed potential antitumor activity with IC_{50} values of 290 μg/disc and >400 μg/disc, respectively (Mazid et al., 2011). This species has also been used in traditional medicines for the treatment of fever, pain, and inflammation.

NUPHAR LUTEA (L) SM.

Phylum: Magnoliophyta
Class: Magnoliopsida
Order: Nymphaeles
Family: Nymphaeceae

Common name: Yellow pond lily

Distribution: Europe, Northwest Africa, Western Asia

Habitat: Shallow waters, wetlands, ponds, slow streams

Description: This species, rooted in sediments, grows in water up to 5 m deep. Its leaves float on the water surface. The solitary terminal and hermaphroditic flower (3–5 cm diameter), which has many yellow petals, is also held above the water surface.

Compounds and activities: The bioactive compounds isolated from this species include thionupharidines, thionuphlutidines, nupharolutine, and 6,6-dihydroxythiobinupharidine. Other compounds of this species are nuphor alkaloids (NUPs) and dimeric sesquiterpene thioalkaloids containing thionupharidines. The NUPs inhibit the NF kappa B pathway and serve as sensitizers to conventional chemotherapy, which may have applications against cancer and inflammation (Ozer et al., 2009). The dimeric sesquiterpene thioalkaloids, however, have shown antibacterial and antileishmanial (against the protozoan parasite, *Leishmania major*) activities. This plant is also used in homeopathic medicines.

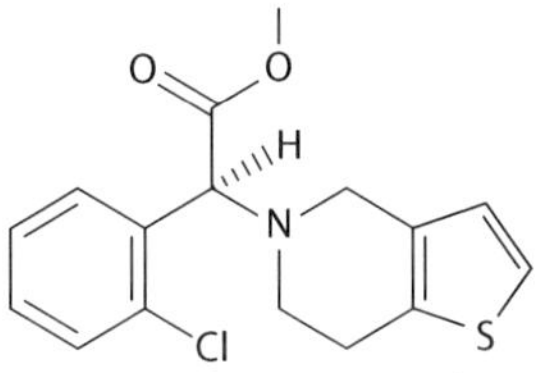

thionupharidines

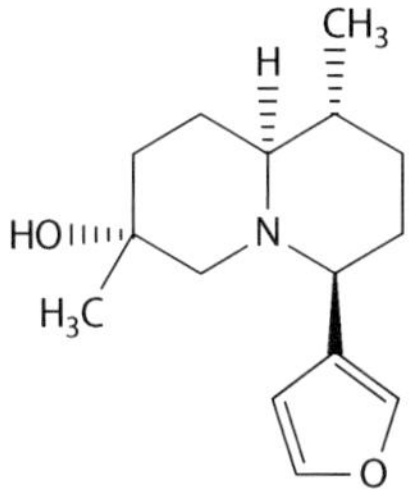

nupharolutine

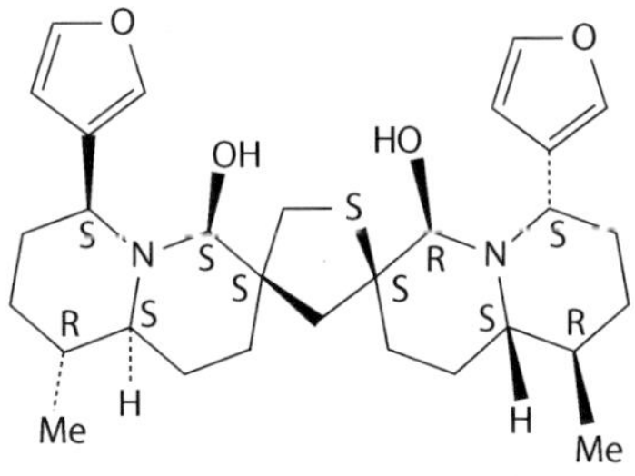

6,6-dihydroxythiobinupharidine

NYMPHAEA PUBESCENS WILLD

Phyllum: Magnoliophyta Class: Magnoliopsida
Order: Nymphaeles Family: Nymphaeceae
Common name: Hairy water lily, pink water lily
Distribution: Temperate and tropical Asia, Bangladesh, India, Pakistan, Sri Lanka, Japan, Philippines, Cambodia, Laos, Myanmar, Thailand, Vietnam, Indonesia, Malaysia, Australia, Papua New Guinea
Habitat: Shallow lakes, ponds
Description: This is an erect, rhizomatous, and perennial species possessing rounded leaves above water and heart-shaped leaves below water. The flowers (15 cm diameter) are white to pink.
Compounds and activities: The ethanolic flower extracts of this species possess anticancer properties and are activated against human cervical carcinoma HeLa cell lines and MCF-7 cell lines. The IC_{50} values for these cell lines were 91.57 and 99.6 μg/ml, respectively (Selvakumari et al., 2012). The flower extracts of this species possess alkaloids, ellagicacid, phenols, tannins, saponins, and flavonoids. The aqueous and ethanol extracts of the tubers of this species possess hypoglycemic, hypolipidemic, and antioxidant activities.

UTRICULARIA AUREA LOUR, 1790

Phylum: Magnoliophyta Class: Magnoliopsida
Order: Lamiales Family: Lentibulariacea
Common name: Golden or leafy bladderwort
Distribution: India, Japan, China, Southeast Asia, Malaysia, Australia
Habitat: Shallow, deep, stagnant or slow-moving freshwaters, ditches, ponds, lakes
Description: It is a partially submersed, insectivorous aquatic plant growing to only less than 50 cm. The simple, sessile, and finely divided leaves are alternate and are spirally coiled along the rounded stem. The axillary or terminal racemose flowers are yellow and bisexual.
Compounds and activities: The bioactive compound fucoidan (a sulfated polysaccharide) of this species possesses antitumor and anticoagulant properties. A quantity of 500 μg/ml of fucoidan was found to inhibit the growth of the epidermoid carcinoma cell and fibroblast cell at 84 and 55%, respectively (Choosawad et al., 2005). The crude fucoidan showed the activated partial thromboplastin time (APTT) at 4.89 ± 0.92 IU/mg.

ROTULA AQUATICA LOUR

Phylum: Magnoliophyta Class: Magnoliopsida
Order: Tubiflorae Family: Boraginaceae
Common name: Aquatic rotula
Distribution: India, China
Habitat: Rock crevices in streambeds
Description: This rare rheophyte is an aromatic shrub (2.5 m height) that has small branches. Leaves (1.5 cm long) are alternate, overlapping, and ovate or oblong. Flowers are small (6 mm long), crowded, and borne in clusters. Each flower is 6 mm long with five purple or pink petals. The fruit (4 mm diameter) is fleshy and rounded.
Compounds and activities: The alcoholic and aqueous extracts of roots of this species contain bioactive compounds such as alkaloids, flavonoids, phenolic compounds, and steroids. Another investigation on the root extracts of this species showed the presence of allantoin and rhabdiol (sterol) as

bioactive compounds. All these compounds are responsible for the antipyretic, analgesic, anti-inflammatory and antihelmintic activities of this species. The polyphenols (tannins) and steroids of this species have shown antimitotic properties by preventing cell proliferation of pancreatic cancer cell lines (Patil et al., 2004). This species has also been traditionally used in the treatment of kidney and bladder stones. Its root tubers have shown diuretic properties and have been used in the treatment of cough, heart disease, dysuria (painful urination), blood disorders, fever, poisonings, ulcer, and uterine diseases (Sivarajan and Balachandran, 1994).

LYSIMACHIA CLETHROIDES DUBY

Phylum: Magnoliophyta
Class: Magnoliopsida
Order: Primulates
Family: Primulaceae
Common name: Gooseneck loose strife
Distribution: China, Japan, Russia, Korea, North America
Habitat: Damp woodland margins, wet ravines
Description: This perennial, herbaceous, flowering species has a maximum height of 1 m, and its stem is rigid and upright. The leaves (10 × 5 cm) are alternate, and oblong or broadly lanceolate with entire margins. The small white flowers (1 cm wide) with five petals are grouped in terminal spikes.
Compounds and activities: The oleanane-type triterpenoid saponins, viz. clethroidosides A–G and declethroidoside H, have been isolated from this species. Of them, declethroidoside H has been found to be cytotoxic against human tumor cell lines (HT-29, He PG 2, BGC-823, A 549, A 375 S180, H22 U14, and L1210) (Liang et al., 2011). The plant extract also reduces cell apoptosis in HL-60, SMMC-7721, and K562 cells, and inhibited metastasis in hepatoma and uterine cervix cancer (Yang et al., 2009).

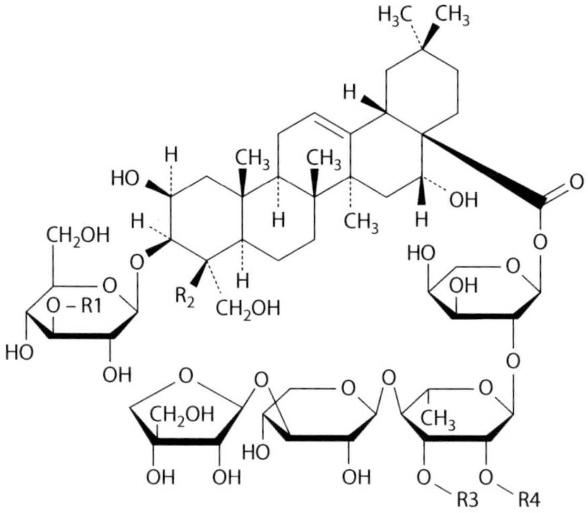

triterpenoid saponins

LYSIMACHIA NUMMULARIA LINNAEUS

Phylum: Magnoliophyta Class: Magnoliopsida
Order: Ericales Family: Myrsinaceae
Common name: Golden creeping fenny, money wort
Distribution: Europe, North America
Habitat: Sides of streams, marshes, ponds
Description: This is a low, creeping, submersed or immersed, golden color plant that forms evergreen ground cover with rounded, ovate, and yellow leaves. Many cup-shaped bright yellow flowers are present.
Compounds and activities: The cytotoxic bioactive compounds isolated from this species include ursane-type triterpenoid saponins, viz., clethroidosides

A–G; triterpenoid saponins, viz., clethroidoside H; and olean-type triterpenoid saponins. In folk medicines, this plant is used for treating wounds and gall and urinary bladder stones and for painful gout symptoms (Yang et al., 2009; Liang et al., 2011).

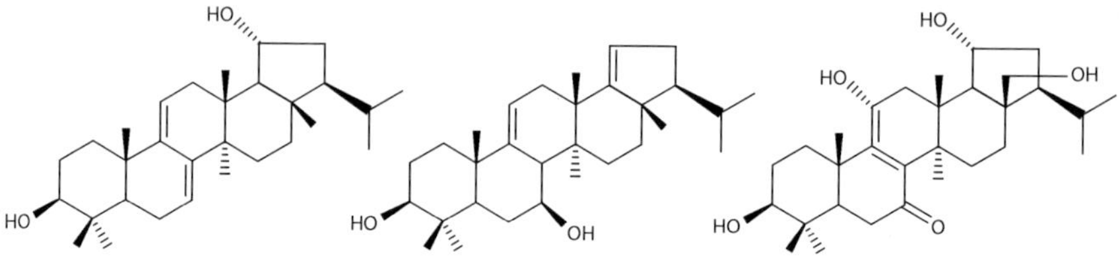

clethroidoside A-H

LYSIMACHIA CONGESTIFLORA HEMSI L

Phylum: Magnoliophyta Class: Magnoliopsida
Order: Ericales Family: Myrsinaceae
Common name: Golden globe, creeping Jenny
Distribution: Southeast and Southwest China
Habitat: Damp and moist places
Description: This prostrate, spreading perennial herb grows to a maximum height of only 20 cm. It has yellow-green foliage. Mid-green leaves are red edged and yellow variegated. Yellow-red flowers are cup shaped and five petaled.
Compounds and activities: A triterpene saponoside, 3-O-{β-D-xylopyronosyl-(1->2)β-D-glucopyronosyl-(1->4)-[β-D-glucopyranosyl-(1->2)]-α-L-arabinopyranosyl}-cyclamiretin A (LTS-4), has been isolated from the underground parts of this species. Methanol extract of this species was found to be most active against target cells (sarcoma XC) with 95% dead cells at 80 μg/ml. This could largely be due to the saponins presently isolated, viz., TS-4, and other triterpene saponosides earlier isolated, viz.,

LTS-1 and LTS-7. The ED_{50} values for these saponins were 4.5, 7.0, and 6.0 μg/ml, respectively. Colechin, which was used as a reference substance, had an ED_{50} of 2.5 μg/ml (Podolak et al., 2007).

JUSSIAEA REPENS LINNAEUS

Phylum: Magnoliophyta Class: Magnoliopsida
Order: Myrtales Family: Onagraceae
Common name: Floating primrose willow
Distribution: United States
Habitat: Shallow waters, pond margins
Description: This herbaceous species is reddish with fibrous roots. The stem is erect to repent on land or floating. It grows to a maximum length of 50 cm. Leaves are alternate, petiolate, and oblong to elliptic with a length of 12 cm and breadth of 2 cm. Single axillary flowers are present on long peduncles. Flowers are yellow with five petals.
Compounds and activities: The bioactive compounds isolated from this species include rosmarinic acid, quercetin-3-O-D-glucopyranoside, and kaempferol-3-O-D-glucopyranoside. Isolated flavonoids of this species have shown cytotoxic activity against Ehrlich ascitis carcinoma cells (Marzonk et al., 2007). Ethyl acetate extract of this species possesses antioxidant, hepato-protective, anti-inflammatory, and antidiabetic properties.

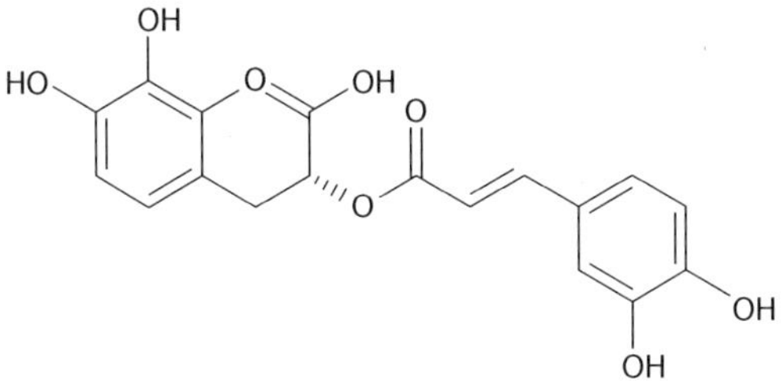

rosmarinic acid

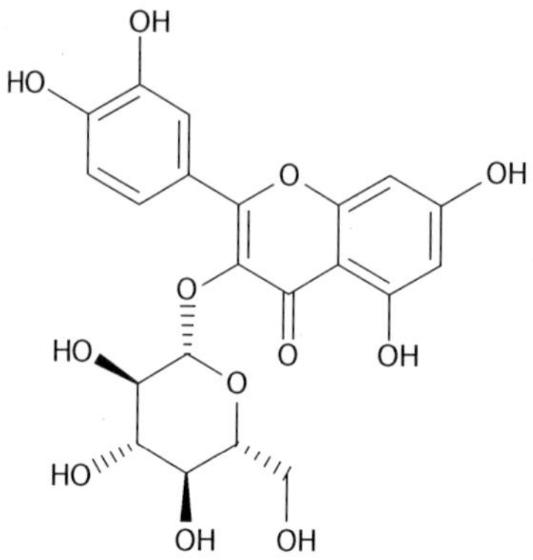

quercetin-3-O-D-glucopyranoside

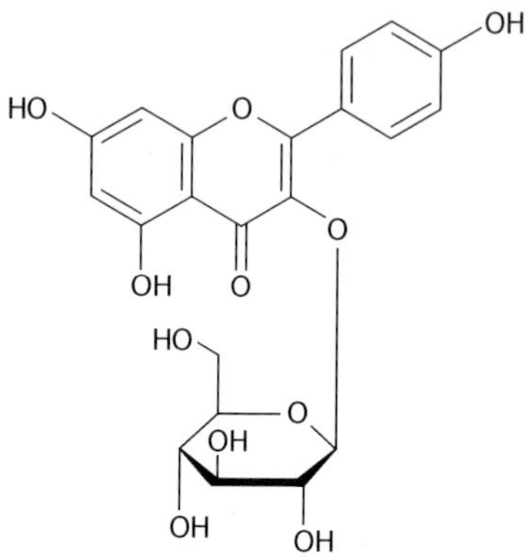

kaempferol-3-O-D-glucopyranoside

LUDWIGIA HYSSOPIFOLIA (G. DON)

Phylum: Magnoliophyta
Class: Magnoliopsida
Order: Myrtales
Family: Onagraceae
Common name: Seed box, primrose willow
Distribution: Africa, Asia, Sri Lanka, India, China
Habitat: Pools, shallow ditches, river edges
Description: This annual herbaceous plant grows to a maximum height of 3 m. The stem is glabrous, branched, and green to red colored with striations. Leaves (10 × 3 cm) are lanceolate to linear-lanceolate and white-green. Flowers are small, axillary, solitary, and sessile with four yellow, white, orange, or red petals that are ovate or elliptic.
Compounds and activities: An alkaloid compound, 1-[5-(1,3-benzodioxol-5-yl)-1-oxo-2,4-pentadienyl] piperidine (piperine), possessing antitumor and antibacterial activities was isolated from this species. Piperine exhibited antitumor activity with an IC_{50} value of 13.50 μg/disc. Ethyl acetate extract has been reported to possess antitumor properties (Das et al., 2007) and showed 73.05 and 84.14% inhibition of *Agrobacterium tumefaciens*-induced crown

gall tumor formation in potato disc. Both ethyl acetate and piperine exhibited moderate antibacterial activity against selected Gram-positive and Gram-negative bacteria (Uddin et al., 2012).

LYTHRUM SALICARIA LINNAEUS

Phylum: Magnoliophyta Class: Magnoliopsida
Order: Myrtales Family: Lythraceae
Common name: Purple loosestrife
Distribution: Europe, Asia, Northwest Africa, Southeast Australia
Habitats: Shallow waters, stream and riverbanks, lake shores, wet meadows, pastures, and marshes
Description: It is a herbaceous, perennial flowering plant species. It grows to 1.5 m, and its stems are reddish, purple, or red to purple. The maximum length and breadth of its lanceolated leaves are 10 cm and 15 mm, respectively. They are sessile and arranged opposite or in whorls of three. The flowers are reddish purple (10–20 mm diameter).
Compounds and activities: The total extract of this species has been reported to possess anticancer properties and inhibit the HT 29 cell line. Similarly, the triterpenoids containing ethyl acetate and chloroform fractions have possessed antineoplastic and cytotoxic effects toward K562 and J47 D cancer cell lines (Khanavi et al., 2011). From the methanol extracts of the seeds, an antibacterial compound, viz., hexahydroxydiphenyl ester (vescalagin), and antifungal compounds, viz., oleanoic acid and ursolic acid, have been isolated. These compounds showed inhibitory action against *Staphylococcus aureus*, *Pseudomonas aeruginosa*, *Escherichia coli*, and *Candida albicans*. Further, human health-promoting flavonoids such as vitexin, isovitexin, orientin, and isoorientin have also been isolated from this species. This astringent medical herb has also been used in traditional medicines for treating diarrhea and dysentery for all age groups, including babies. The antioxidant activity of the seeds of this species has been tested using trolox extract (TE), and the value recorded was 261,384 μm TE (Borchardt et al., 2009).

ALTERNANTHERA PHILOXEROIDES GRISEB

Phylum: Magnoliophyta
Class: Magnoliopsida
Order: Caryophyllales
Family: Amaranthaceae
Common name: Alligator weed
Distribution: South America, Australia, China, New Zealand, Thailand, United States
Habitat: Freshwater systems
Description: This species grows to a maximum size of 10 m. Stems of this plant are long, branched hollow, fleshy, and succulent. Leaves are simple, elliptic with smooth margins, and are opposite in pairs or whorls. Flowers (10 cm) are whitish and ball shaped.
Compounds and activities: From the aerial parts of this species, alternananthin B and N-trans-feruloyl-3,5-dimethoxytyramine possessing antitumor properties have been isolated (Fang et al., 2007).

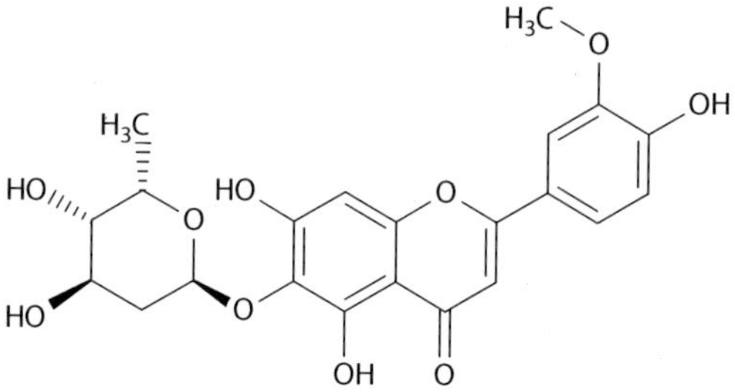

alternananthin B

PFAFFIA GLOMERATA (SPRENG) PEDERSEN (= *GOMPHRENA STENOPHYLLA*)

Phylum: Magnoliophyta
Class: Magnoliophyceae
Order: Caryophyllales
Family: Amaranthaceae
Common name: Brazilian ginseng
Distribution: Argentina, Bolivia, Brazil
Habitat: Rivers
Description: This species grows to a maximum height of 2 m and has hairy leaves.
Compounds and activities: The oleanone-type triterpenes isolated from this species include oleanolic acid, (23 Z)-feruloyl hederagenin, (23 E)-feruloyl hederagenin, β-amyrin acetate, and β-amyrin palmitate. The root extract of this species showed cytotoxicity, and the IC_{50} value was found to be >2,000 μg/ml (Gosmann et al., 2003). The compound β-ecdysone isolated from the roots of this species possesses antidiabetic and aphrodisiac properties. While the methanolic extract of this species possesses free radical scavenging activity and has cytotoxic activity, its butanolic root extract, which has 20-hydroxy ecdysone, possesses free radical scavenging activity. The hydroalcoholic extract of this species was found to be relatively active against *Leishmania braziliensis*.

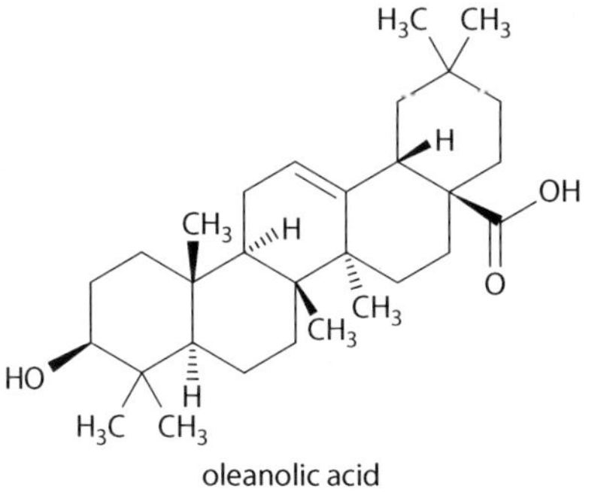

oleanolic acid

IPOMOEA AQUATICA FORSSK

Phylum: Magnoliophyta
Class: Magnoliopsida
Order: Solanales
Family: Convolvulaceae
Common name: Swamp morning glory, water spinach, river spinach
Distribution: Tropical and subtropical regions of world
Habitat: Freshwater swamps, waterways, semiaquatic habitats
Description: This perennial species is a floating herbaceous vine and has stems measuring 2–3 m. Its leaves (15 × 8 cm) are arrowhead shaped or lanceolated. The flowers, measuring 5 cm in diameter, are white and trumpet shaped.
Compounds and activities: The extracts of this species possess antiproliferative, antioxidative, antimicrobial, and antidiabetic activities. The antiproliferative activities of the extracts of this species studied showed that water extract of the stem had the highest antiproliferative activity, with EC_{50} values of 661.40 ± 3.36 μg dry matter/ml, followed by ethanol extract of the stem and ethanol extract of the leaf. The water extract of the leaf had the lowest antiproliferative activity (EC_{50} > 1,000 μg dry matter/ml) (Huang et al., 2005). It was also found that the ethanol extract of the stem of this species showed the highest radical scavenging activity, followed by ethanol extract of the leaf. Its bioactive compound, 7-O-β-D-glucopyranosyl-dihydroquercetin-3-O-α-D-glucopyranoside, is a promising agent for the production of several medicinally important compounds, especially for the treatment of high blood pressure, peptic ulcers, and eye diseases. The phenolic compounds of this species have shown antioxidant and disease-preventing characteristics. The carotenoids of this species, viz., lutein, lutein epoxide, violaxanthin, and neoxanthin, possess antioxidant, anticancer, and anti-inflammatory activities. Further, its extract has also been reported to possess antidiabetic, antimicrobial, and antiulcerogenic properties. The role of this plant in improving the acetylcholine level in the brain and subsequent prevention of memory loss and Alzheimer's disease is worth mentioning.

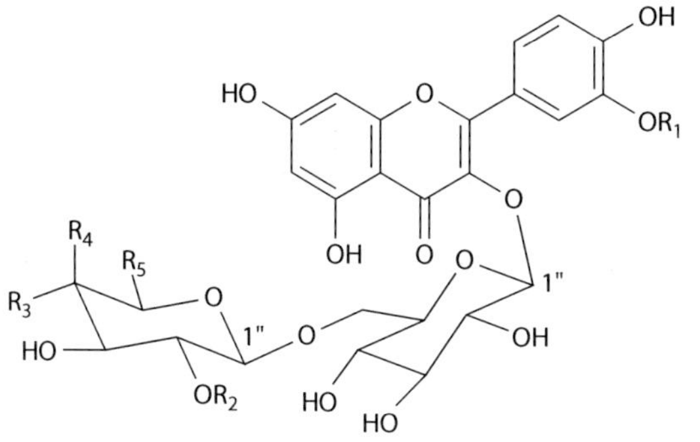

R_1 = Me, R_2 = Ac, R_3 = OH, R_4 = H, R_5 = CH_2OH

7-O-β-D-glucopyranosyl-dihydroquercetin-3-O-α-D-glucopyranoside

POLYGONUM MINUS HUDSON, 1762

Phylum: Magnoliophyta Class: Magnolopsida
Order: Polygonales Family: Polygonaceae
Common name: Knotweed, knotgrass
Distribution: Europe, Scandinavia, Spain, temperate Asia
Habitat: Wet marshy places and beside lakes and ponds
Description: This annual herb grows to a maximum height of 0.3 m. The stem is very slender, stout, or glabrous. Leaves (12 cm long) are shortly petiolate, lanceolated, or linear. Inflorescence (6 cm long) is with six flowers. Flowers are hermaphroditic.
Compounds and activities: The ether extracts of this species contain a flavone, viz., 6,7-methylenedioxy-5,3′,4′,5′-tetramethoxyflavone, and a methyl flavonol, viz., 6,7,4′5′-dimethylene dioxy-3,5,3′-trimethoxy flavone. The cytotoxicity of this species against HeLa cells has been demonstrated with a CD_{50} value of 0.1 mg/ml (Qadar et al., 2012). The extracts of this plant have also shown antioxidant, antimicrobial, and antiulcer activities.

POLYGONUM CUSPIDATUM SICBOLD & ZUCC.

Phylum: Magnoliophyta
Class: Magnoliopsida
Order: Polygonales
Family: Polygonaceae
Common name: Japanese knotweed
Distribution: Japan, China, Korea, North America, Europe
Habitat: Along streams and rivers
Description: It is a perennial species with spreading rhizomes and many reddish brown, freely branched stems. This shrub may grow to a maximum height of 2.5 m. Leaves (15 cm long) are ovate and petiolated. Whitish dioecious flowers are produced in erect racemes of 15 cm long.
Compounds and activities: Four anthroquinones, viz., physcion, emodin, citreorosein, and anthraglycoside B, and two stilbenes, viz., resveratrol and piceid, were isolated from the root extracts of this species. Resveratrol isolated from the root of this species, at doses 2.5 and 10 mg/kg, significantly reduced the tumor volume (42%), tumor weight (44%), and metastasis to the lung (56%) in mice bearing highly metastatic Lewis lung carcinoma (LLC) tumors (Kimura and Okuda, 2001). Further, this compound inhibited DNA synthesis significantly in LLC cells, and its inhibitory concentration (IC_{50}) was 6.8 μmol/L. Furthermore, this compound at 100 μmol/L increased apoptosis to 20.6 ± 1.35% from 12.1 ± 0.36% ($p < 0.05$) in LLC cells. Resveratrol inhibited tumor-induced neovascularization at doses of 2.5 and 10 mg/kg in an *in vivo* model (Kimura and Okuda, 2001). Compounds physcion, emodin, citreorosein, and anthraglycoside B showed moderate to strong inhibition of tyrosinase, and they may be useful as skin-whitening agents to inhibit tyrosinase for dermal use (Leu et al., 2008). Further, certain compounds isolated from this species have also been reported to possess anti-HIV properties (Lin et al., 2010).

ACORUS GRAMINEUS SOLAND AITON

Phylum: Magnoliopsida
Class: Liliopsida
Order: Acorales
Family: Acoraceae
Common name: Japanese sweat flag
Distribution: Eastern Asia, China, Japan, India
Habitat: Wetlands, shallow waters, edges of ponds and streams
Description: It is a fully or partly submerged perennial shrub with long, narrow, and slightly curved golden yellow leaves, small yellow hermaphroditic flowers, reddish fleshy fruits, and fleshy rhizome. It grows to a maximum height of 30 cm only.
Compounds and activities: Bioactive compounds isolated from the rhizome extracts of this species include asarones, 1,2-dimethoxy-4-(E-3′-methyl oxiranyl) benzene, 1,2,4-trimethoxy-5-(Z-1-propenyl)benzene, 1,2,4-trimethoxy-5-(E-3′-methoxyiranyl) benzene ligraminols A–E, and two new 8-O-4′-neolignans, viz., surinamensinols A and B and acoromol. Among them, surinamensinols A and B showed moderate antiproliferative activities against A549, SK-OV-3, SK-MEL-2, and HCT-15 cell lines with IC_{50} values in the range of 4.17–26.18 μm (Kim et al., 2012). Ligraminols A–E have been found to be cytotoxic against human cancer cell lines. Asarone has antiepileptic and antibacterial activities. The root extracts of this plant have antirheumatic, antihelmintic, antiperiodic, antispasmodic, sedative, and vermifuge properties besides treating diarrhea, cough, bronchial asthma, and neurasthenia (nerve weakness).

CAREX VULPINOIDEA MICHX

Phylum: Magnoliophyta Class: Liliopsida
Order: Cyperales Family: Cyperaceae
Common name: Fox sedge
Distribution: North America, Europe, New Zealand
Habitat: Wet habitats, moist to wet shores, meadows, along banks of streams or ponds
Description: This species produces clumps of stems that grow to 1 m. Inflorescence is a dense, tangled cluster of many flower spikes up to 10 cm long. Foliage and flowers are green in color.
Compounds and activities: The bioactive compounds isolated from the seeds of this species are hopeaphenol (resveratrol tetramer), luteolin, 3,5,5′,7′-tetrahydroxy flavonone, and methylated naringenin (polyphenols). Among them, the compound hopeaphenol was found to possess cytotoxicity against human colon cancer (HCT-116, Caco-2), and the IC_{50} value was found to be 2–5 μm (Niesen et al., 2012).

7 Antioxidant Compounds

The oxidative stress that is induced by oxygen radicals is largely responsible for various degenerative diseases of the central nervous system (CNS), as well as the normal process of aging. Protection from aging, mutagenesis, carcinogenesis, etc., is mainly due to antioxidant effects. The antioxidants or radical scavengers are compounds that inhibit or delay oxidation of other molecules. Natural antioxidants include phenolic compounds, nitrogen compounds, vitamin C (ascorbic acid), vitamin E (tocopherols), flavonoids, phenolic acids, carotenoids, etc.

CLADOPHORA GLOMERATA (LINNAEUS) KUTZING, 1843

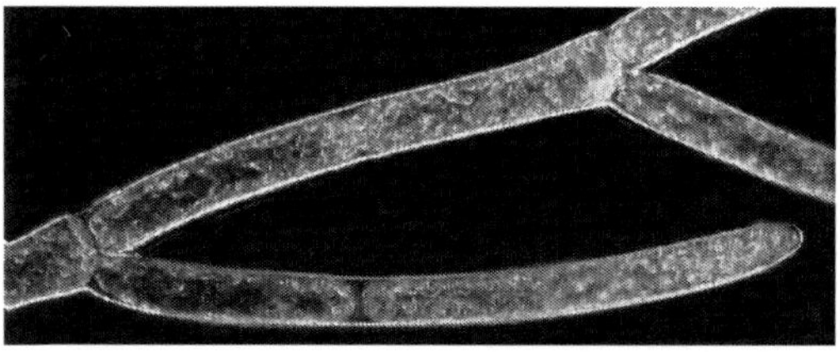

Phylum: Chlorophyta
Class: Chlorophyceae
Order: Cladophorales
Family: Cladophoraceae
Distribution: Throughout the world's freshwater ecosystems, Europe, North, Central, and South America, Southwest and Southeast Asia
Habitat: Rivers, streams, lakes
Description: This singly filamentous species is a rapid colonizer of rocky streambeds. The thalli of uniseriate dark green filaments (200 cm long) are provided with apical and intercalary sheaths. The cells (100 × 100 μm) are multinucleate and cylindrical and have parietal chloroplasts. The apical cells (1,100 × 90 μm) are cylindrical and fairly tapering. The akinetes are formed in this species only during unfavorable conditions.
Compounds and activities: The crude methanol extract of the species has shown antioxidant, antibacterial, and antiprotozoal activities. The values of total phenolic and flavonoid contents of the extract of this species were 3,077 ± 105 mg gallic acid equivalent/g and 595 ± 23 mg quercetin equivalent/g. The scavenging activity of 2,2-diphenyl-1-picrylhydrazyl (DPPH) radicals (920 ± 42 μg/ml IC_{50} value), iron chelating activity (IC_{50} = 971 ± 39 μg/ml), and scavenging activity of nitric oxide (0.20 ± 0.01 mg/ml IC_{50}) have been recorded in this species (Soltani et al., 2011).

PEDIASTRUM DUPLEX MEYEN, 1829

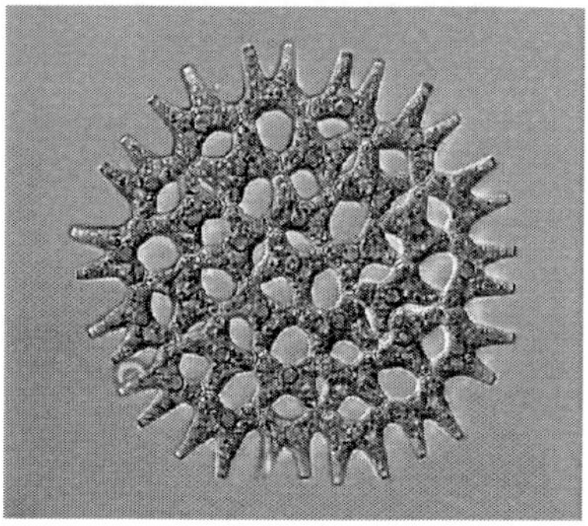

Phylum: Chlorophyta Class: Chlorophyceae
Order: Sphaeropleales Family: Hydrodictyaceae
Distribution: Europe, North America, South America, Asia, Southeast Asia, Australia, New Zealand, Pacific Islands, Caribbean Islands
Habitat: Freshwater habitats
Description: This species forms nonmotile coenobia (colonies) of a varied number of cells, from 5 to 130. The cells are flat and circular shaped. The cell bodies are polygonal and are granulated with horn-like projections. The reproduction is by asexual methods, i.e., by producing autocolonies.
Compounds and activities: All termamyl (carbohydrase) and kojizyme (protease) digests of this species showed significant 1,1-diphenyl-2-picrylhydrazyl (DPPH) scavenging effects and both no scavenging and metal chelating. Termamyl (60.6%) digest of this species possessed the highest effects on hydrogen peroxide scavenging. Lipid peroxidation was significantly inhibited in viscozyme. The water-soluble antioxidants of this species may be applied in pharmaceutical industries in the production of novel drugs (Lee et al., 2009).

SPONGIOCHLORIS SPONGIOSA (VISCHER) STARR, 1955

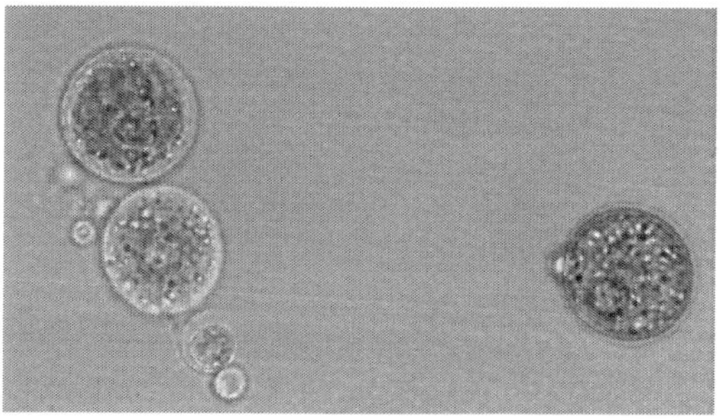

Phylum: Chlorophyta Class: Chlorophyceae
Order: Chlorococcales Family: Chlorococcaceae

Distribution: Europe, Southeast Asia
Habitat: Freshwater systems
Description: The cells of this species are spherical or subspherical with cup-shaped chloroplast and one pyrenoid. Biflagellated zoospores are produced by this species.
Compounds and activities: Bioactive phenolic acid compounds (protocatechuic, p-hydroxybenzoic, 2,3-dihydroxybenzoic, chlorogenic, vanillic, caffeic, p-coumaric, and salicylic acids), cinnamic acid, and hydroxybenzaldehydes (p-hydroxybenzaldehyde, 3,4-dihydroxybenzaldehyde, and vanillin) possessing antioxidative and anticancer properties have been isolated from this species (Onofrejova et al., 2010).

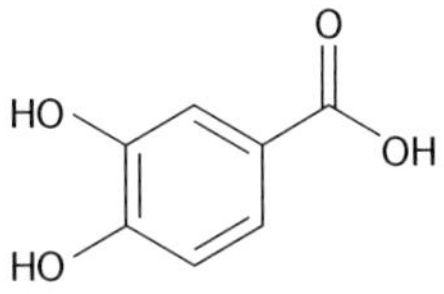
protocatechuic acid

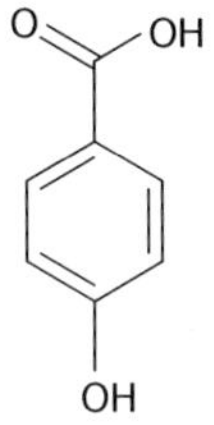
p-hydroxybenzoic acid

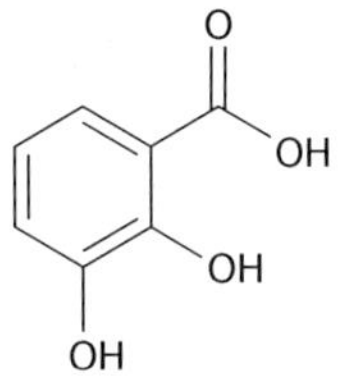
2,3-dihydroxybenzoic acid

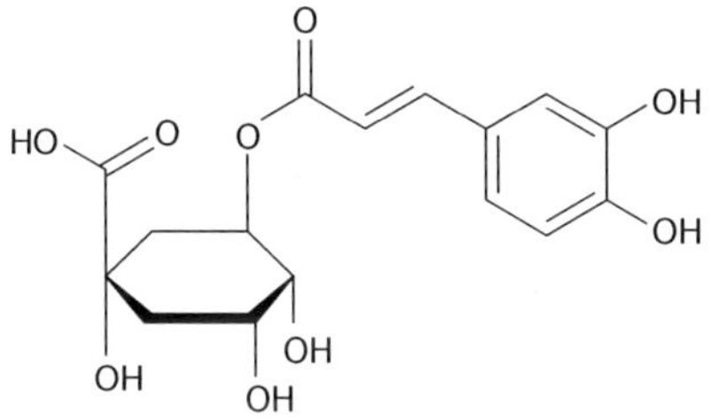
chlorogenic acid

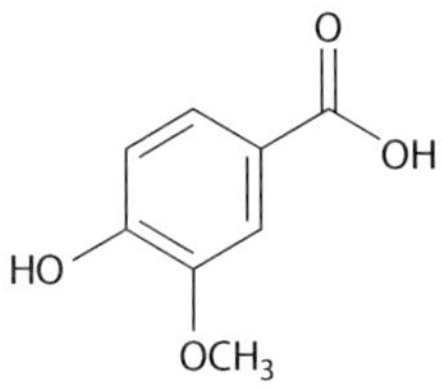
vanillic acid

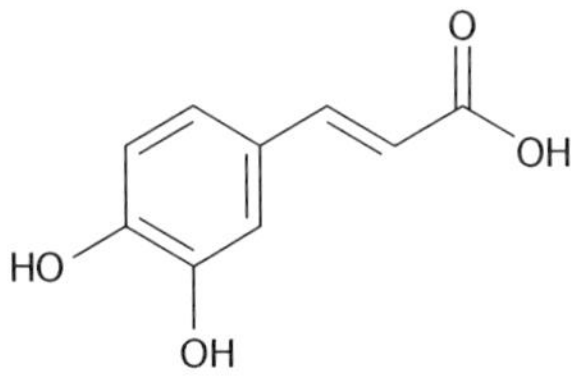
caffeic acid

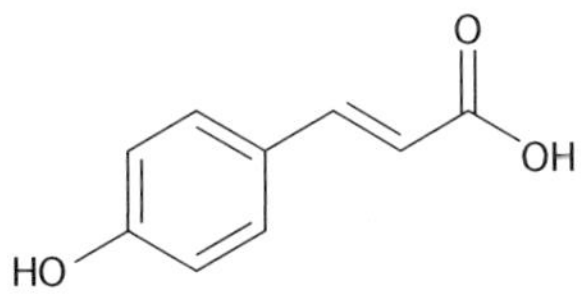
p-coumaric acid

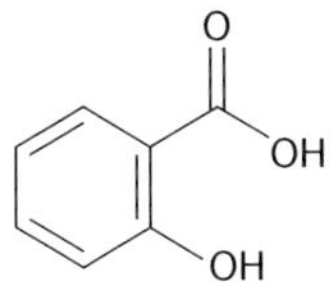
salicylic acid

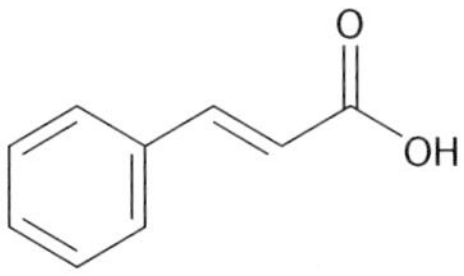
cinnamic acid

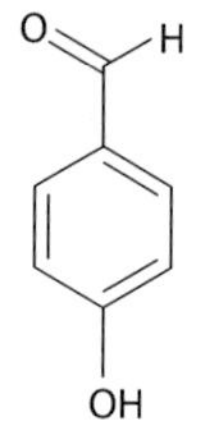

p-hydroxybenzaldhyde

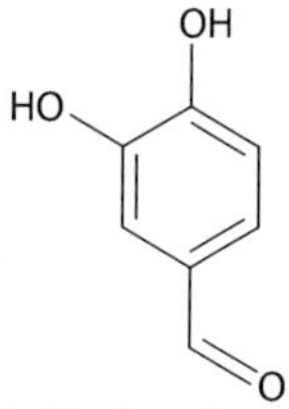

3,4-dihydroxybenzaldehyde

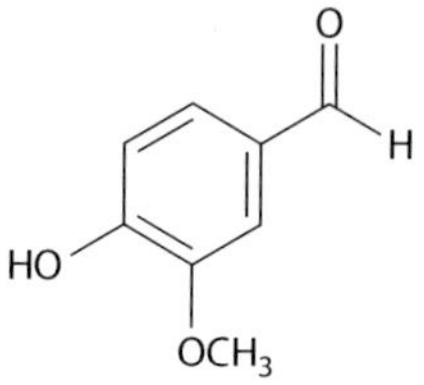

vanillin

DACTYLOCOCCOPSIS FASCICULARIS LEMMERMANN, 1898

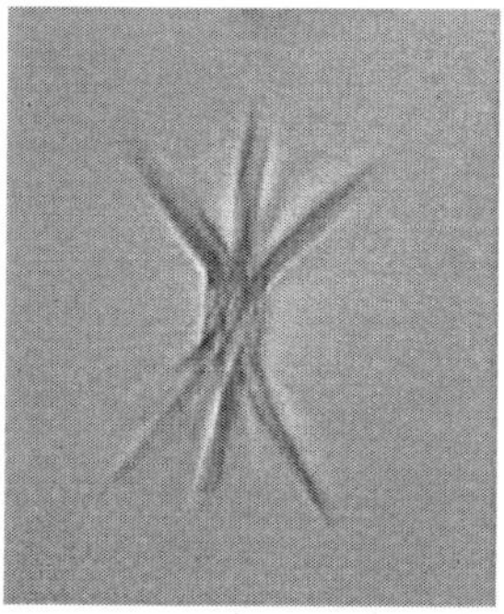

Phylum: Cyanobacteria
Class: Cyanophyceae
Order: Chroococcales
Family: Chroococcaceae
Distribution: Europe, Asia
Habitat: Freshwater habitats
Description: The tiny cells of this species, which are elliptic or fusiform, measure 1–2 μm and 50 μm in width and length, respectively. The cells may be connected by lateral walls or form a branching chain.
Compounds and activities: Celluloclast and kojizyme (carbohydrases) digests of this species showed higher effects on superoxide anion radical scavenging, and the values were 58.1 and 56.9%, respectively. Lipid peroxidation was significantly inhibited in ultraflo, (carbohydrase) protamex, kojizyme, and alcalase (proteases) digests of this species (Lee et al., 2009). The water-soluble, antioxidative polyphenols and polysaccharides of this species have also been reported to be potential sources for producing valuable drugs in pharmaceutical industries.

ANABAENA DOLIOLUM BHARADWAJA, 1935

Phylum: Cyanobacteria
Class: Cyanophyceae
Order: Nostocales
Family: Nostocaceae
Distribution: India
Habitat: Freshwater habitats, dams
Description: This filamentous and atmospheric nitrogen-fixing species has a symbiotic relationship with some plants.

Compounds and activities: The bioactive compounds isolated from this species include phenolic acids (protocatechnic, p-hydroxybenzoic, 2,3-dihydroxybenzoic, chlorogenic, vanillic, caffeic, p-coumaric, salicylic acid), cinnamic acid, hydroxybenzaldehydes (p-hydroxybenzaldehyde, 3,4-dehydroxybenzaldehyde) and vanillin. These compounds possess antioxidative properties and have several nutritional and therapeutic applications (Onofrejova et al., 2010).

NOSTOCHOPSIS LOBATUS WOOD EM GEITLER

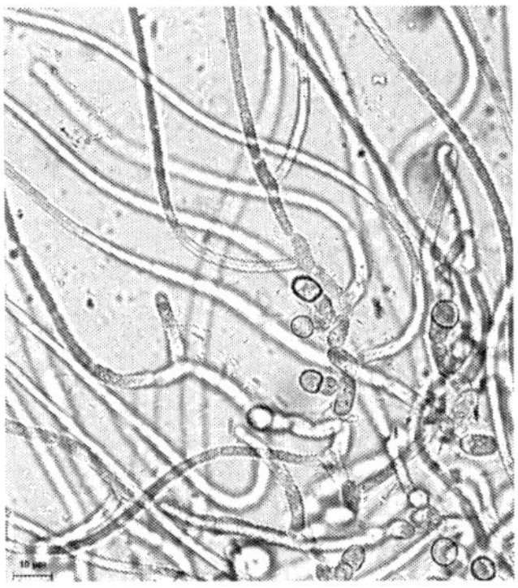

Phylum: Cyanobacteria
Class: Cyanophyceae
Order: Nostocales
Family: Nostocaceae
Distribution: Bangladesh
Habitat: Shallow rivers, streams
Description: The dark green colonies of this species, which are found attached to rocks, possess a mucilagionous, polysaccharide sheath as an outgrowth.
Compounds and activities: The bioactive compounds isolated from this species include δ-aminolevulinic acid, bili pigments, and antioxidants. This species shows a high polysaccharide content of 160.02 DP on BG11 medium, and the values of carotenoid, phycocyanin, and allophycocyanin pigments of this species were found to be 0.339, 61.58, and 65.38 mg/g cell (dw), respectively. This species also showed a high concentration (37 μg/100 g dw) of selenium, a well-known oxidant (Thiamdao et al., 2012). The polysaccharides, phycocyanin and allophycocyanin, of this species could therefore be a source of supplemental food and are used as therapeutic agents and ingredients of cosmetics.

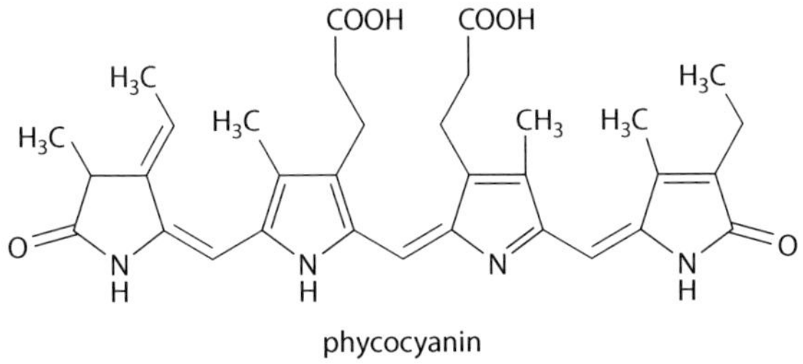

phycocyanin

FALLOPIA SACHALINENSIS F. SCHMIDT (= *POLYGONUM SACHALINENSE*)

Phylum: Magnoliophyta
Class: Magnoliopsida
Order: Caryophyllales
Family: Polygonaceae
Common name: Giant knotweed
Distribution: East Asia, Russia, Britain
Habitat: Riverbanks, along ravines, streams
Description: This herbaceous perennial plant grows to a maximum height of 4 m and has strong, spreading rhizomes. Leaves (40 × 30 cm) are heart shaped, wavy, and with crenate margins. Flowers are small and dioecious and are present in short panicles (10 cm long).
Compounds and activities: The bioactive compounds isolated from this species include phenyl acetonitrile (E)-β-ocimene, linalool, (E)-4,8-dimethyl-1,3,7-nonatriene, and (E,E)-α-farnesene. The polysaccharides of this species have displayed significant radical scavenging activities in the 1,1-diphenyl-2-picrylhydrazyl free radical assay, and they may have potential applications as novel natural antioxidants (Hromadkova et al., 2010). The anthroquinones isolated from the leaves of this species could serve as lyrosinase inhibitors and in the preparation of dermal drugs. Further, the extract of this species has shown antitussive activity.

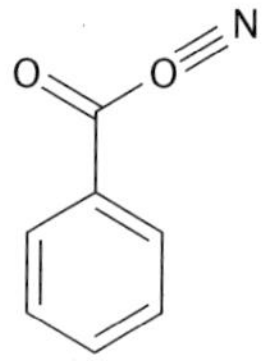

phenyl acetonitrile

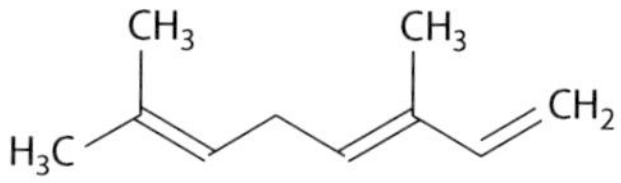

(E)-β-Ocimene

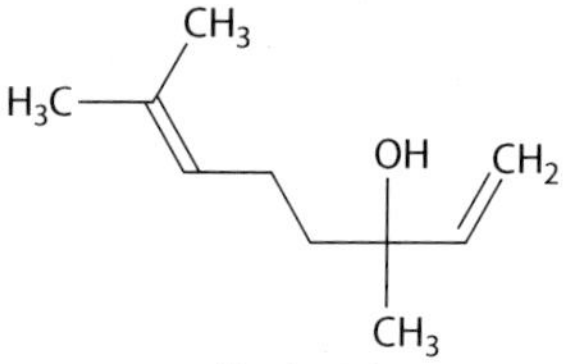

linalool

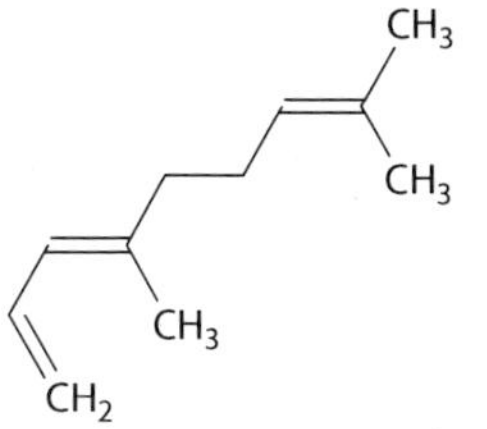

(E)-4, 8-dimethyl-1,3,7-nonatriene

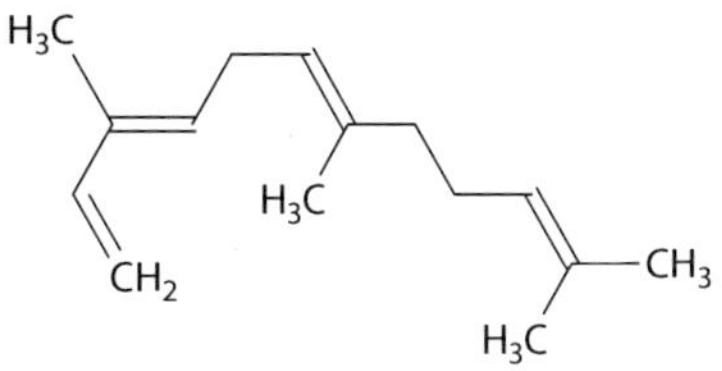

(E,E)-α-farnesene

RUMEX PATIENTIA LINNAEUS

Phylum: Magnoliophyta Class: Magnoliopsida
Order: Caryophyllales Family: Polygonaceae
Common name: Patience dock, garden patience
Distribution: Europe, Russian Federation, Bulgaria, Serbia
Habitat: Meadows, edges of rivers, wet soils
Description: The stem of this species, measuring 80–100 cm in length, is straight and thick with grooves. While the bottom leaves are ovate and measure 20–30 cm long and 8–10 cm wide, the top leaves are lanceolated. The flowers are in whorls, and each whorl consists of 10–15 light brown flowers.
Compounds and activities: The root extracts of this species have been reported to contain antioxidative flavonoids and anthroquinones such as 5-methoxy-7-hydroxy-1(3A)-benzofuronone, 5,7-dihydroxy-1-(34)benzofuronone, adecanic acid-2,3-dihydroxypropyl ester, torachrysone-8-O-β-D-glucopyranoside, gallic acid, β-sitosterol, β-sitosterol-3-O-β-D-glucopyranoside, and catechin. The roots of this species have been used extensively in traditional medicine worldwide for their laxative, diuretic, antipyretic, wound care, and anti-inflammatory properties (Jovin et al., 2011).

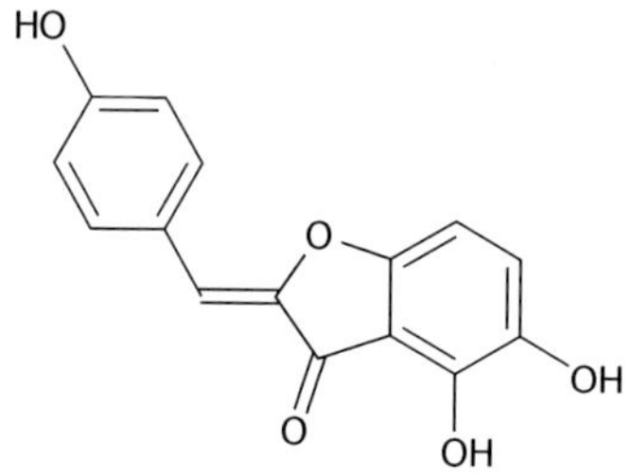

5,7-dihydroxy-1-(34)benzofuranone

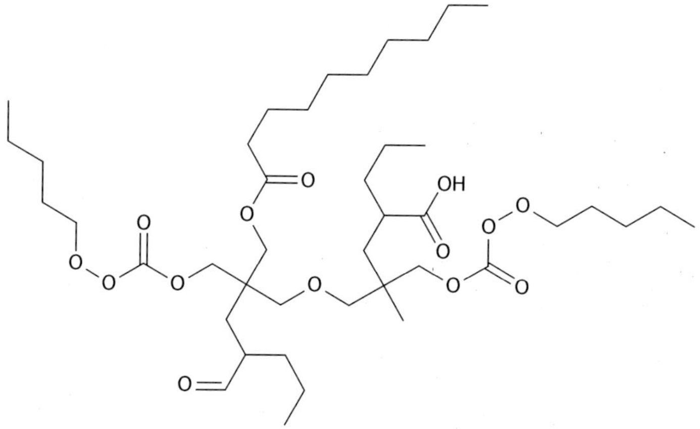

adecanic acid -2, 3-dihydroxypropyl ester

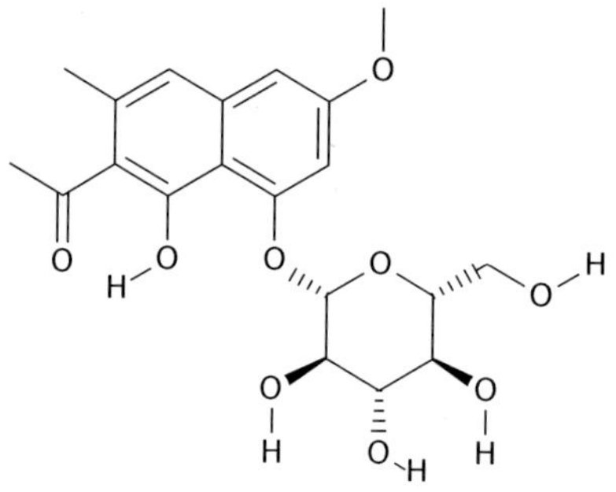

torachrysone-8-O-beta-D-glucopyranoside

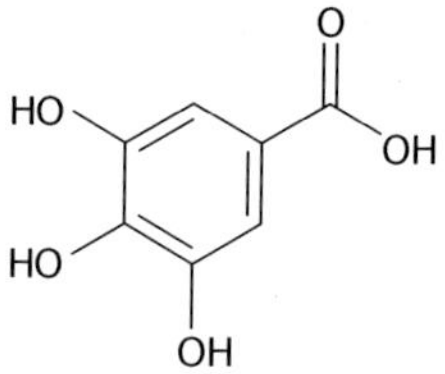

gallic acid

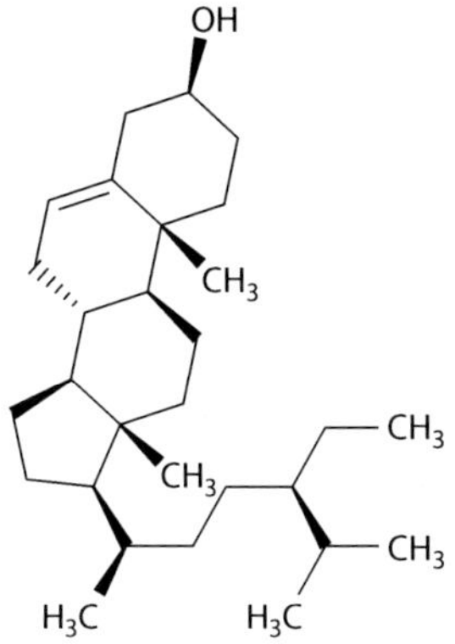

beta-sitosterol

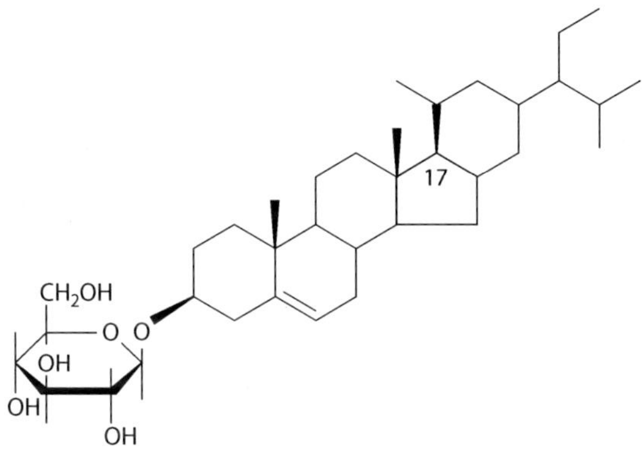

beta-sitosterol-3-O-beta-D-glucopyranoside

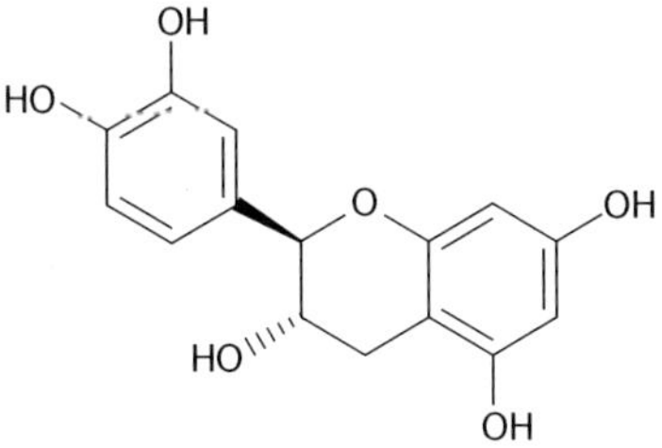

catechin

RUMEX DENTATUS LINNAEUS

Phylum: Magnoliophyta
Class: Magnoliopsida
Order: Caryophyllales
Famiy: Polygonaceae
Common name: Toothed dock
Distribution: Eurasia, North Africa
Habitat: Lake shores, edges of cultivated fields, moist valleys
Description: This annual or biennial herbal species produces a slender and erect stem of 60–80 cm height. The leaves (13 cm) are either oval or lance-shaped with wavy margins. The inflorescence has a cluster of 15–25 flowers.
Compounds and activities: The methanol extracts of this species possess flavonoids, alkaloids, saponins, anthroquinones, and tannins, viz., kaempferol-3-O-β-galactoside, kaempferol-3-O-β-glucoside, kaempferol-3-O-β-rutinoside, isorhamnetin 3-O-β-galactoside, isorhamnetin 3-O-β-glucoside, and isorhamnetin 3-O-β-rutinoside. These bioactive compounds showed antioxidant and antimicrobial activities besides serving as chemotherapeutic agents. The total phenolic content in the ethyl acetate fractions of leaves and roots of this species was high, and the values were 169.5 and 257.4 mg gallic acid equivalent/g extract, respectively. The ethyl acetate fractions of leaves and roots exhibited strong DPPH activity, and the DPPH IC_{50} values were 0.021 and 0.012 mg/ml of leaves and roots, respectively (Elzaawely and Tawata, 2012). Owing to its high levels of phenolic compounds and associated strong antioxidant properties, this species could be utilized as a natural source of antioxidant in the food industry.

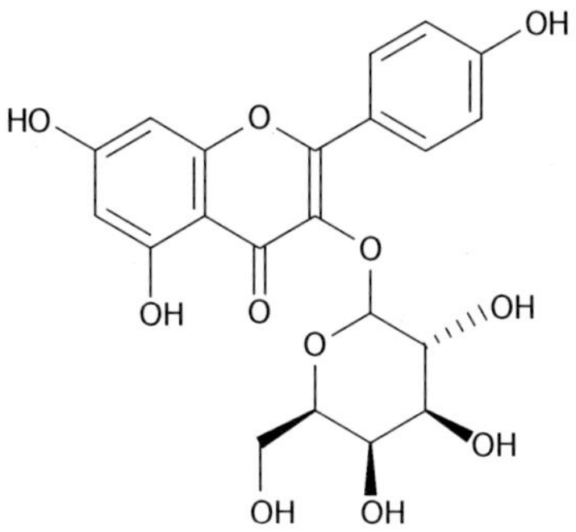

kaempferol-3-O-β-galactoside

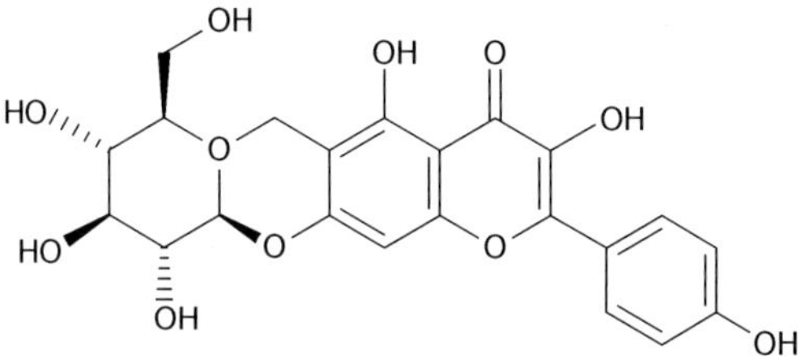

kaempferol-3-O-β-glucoside

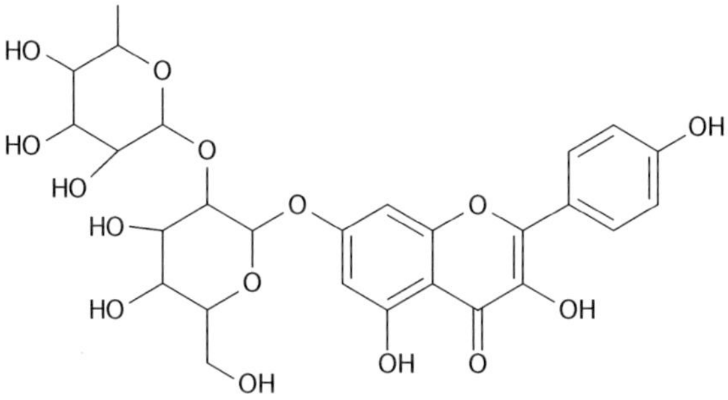

kaempferol-3-O-β-rutinoside

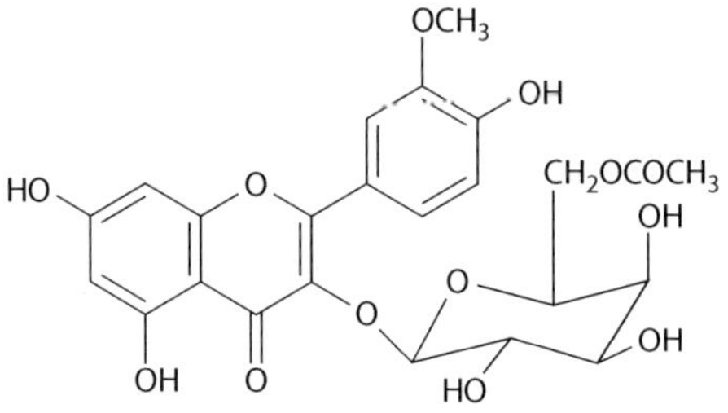

1

isorhamnetin 3-O-β-galactoside

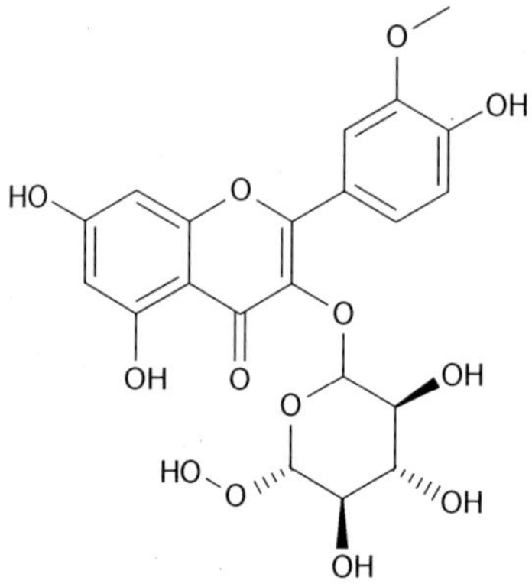
isorhamnetin 3-O-β-glucoside

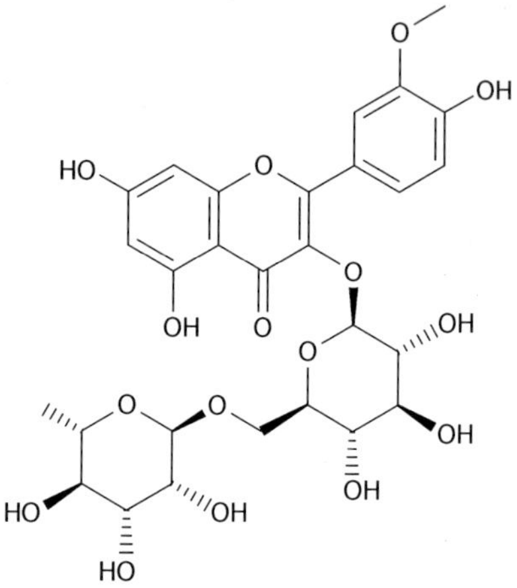
isorhamnetin 3-O-β-rutinoside

RUMEX GMELINI TURTCZ

Phylum: Magnoliophyta
Class: Magnoliopsida
Order: Caryophyllales
Family: Polygenaceae
Common name: Dock, sorrel
Distribution: Britain
Habitat: Stream banks, wet habitats, moist valleys
Description: This perennial herbal species grows to a maximum height of 1 m, and its flowers are hermaphroditic.
Compounds and activities: Two new anthroquinones have been isolated from this species (Kang et al., 1996). The root and rhizome extracts of this species contain bioactive compounds, viz., 1-O-β-D-glucopyranosyl chrysophanol, 1-O-s-chrysophanol, neopodin, emodin, 6-methoxyl-10-hydroxyaolins A and B, and resveratrol, which possess antioxidant, antifungal, antitumor, antitussive, antiasthma, and antiviral activities (Cui et al., 2005).

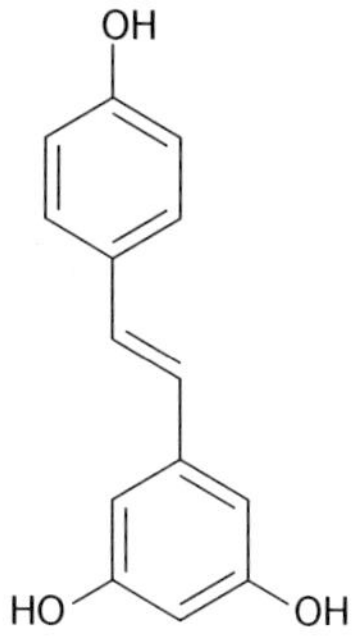

resveratrol

TAMARIX RAMOSISSIMA LEDEB

Phylum: Magnoliophyta
Class: Magnoliopsida
Order: Violales
Family: Tamaricaceae
Common name: Salt cedar
Distribution: Europe, Asia
Habitat: Riparian habitats, water courses, wetlands, estuaries
Description: It is a deciduous shrub or a small tree with reddish stems. Leaves are feathery and plumes of flowers are showy. It grows to maximum height of 8 m with a width of 5 m. This species produces upright racemes of small, pink, five-petaled flowers.
Compounds and activities: The bioactive compound isolated from this species is tamarixetin. The polyphenolic substances isolated from the ethyl acetate and water-acetone extracts of this species have shown antioxidant and antibacterial activities (Sultanova et al., 2001).

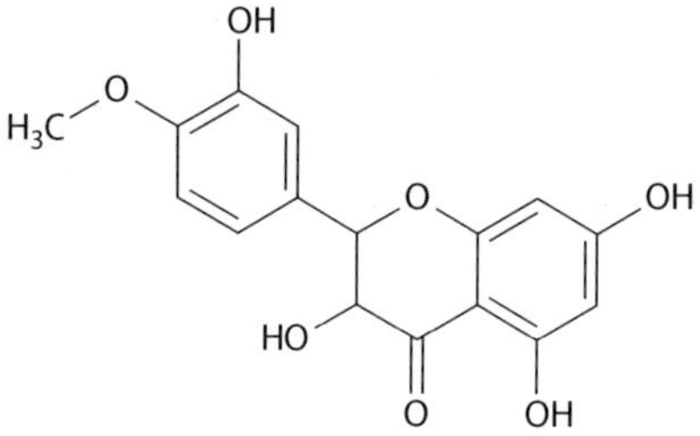

tamarixetin

RANUNCULUS RIONII LAGGER

Phylum: Magnoliophyta
Class: Magnoliopsida
Order: Ranunculales
Family: Ranunculaceae
Common name: Buttercup
Distribution: China, Afghanistan, Pakistan, South Africa, West Asia, Europe
Habitat: Lakes
Description: This annual herb has a branched, glabrous stem of 20 cm length. The leaf blade is flagellate in outline, and flowers (1 cm diameter) have small petals (4 × 2 mm) that are white and ovate.
Compounds and activities: The ethanol extracts of this plant species have shown antioxidant and anticholinesterase (AChE) properties. The total phenolic content of this species was found to be 66.9 ± 0.01 mg/g extract. With regard to the antioxidant property of this species, the values of percentage of inhibition ± SEM against the DPPH radical were 6.3 ± 0.41, 6.4 ± 1.39, 20.3 ± 0.84, 40.4 ±1.27, and 68.3 ± 1.02 for extract concentrations of 0.125, 0.25, 0.5, 1.0, and 2.0 mg/ml, respectively. With regard to the AChE inhibition of the extracts of this species, the values of percentage inhibition ± SEM against AChE were 7.3 ± 0.06 and 8.1 ± 2.09 for the extract concentrations of 1 and 2 mg/ml, respectively (Kartal et al., 2009).

CALTHA PALUSTRIS LINNAEUS

Phylum: Magnoliophyta Class: Magnoliopsida
Order: Ranunculales Family: Ranunculaceae
Common name: King cup, yellow marsh marigold, American cowslip
Distribution: Western temperate countries, United States
Habitat: Marshes, fens, ditches
Description: This herbaceous perennial and polymorphic species has hollow stems and grows to a maximum height of 80 cm, and its leaves (4–20 cm) are rounded, heart shaped, or kidney shaped. The clusters of flowers measuring 2–6 cm diameter are yellow and buttercup-like.
Compounds and activities: The bioactive compounds isolated from this species are protoanemonin and pyrrolizidine. The free radical scavenging activity (IC_{50}) of total methanol extract of this species was found to be 189 ± 98.60 μg/ml (Nikolova et al., 2011). From the root extracts of this species, triterpenoid glycosides possessing anti-inflammatory activities have been isolated. These compounds have also been reported to reduce serum cholesterol and total protein. The flower extracts yield bioactive compounds such as flavonoids, rhamnosides, glucosides, and glucorhamnosides of kaempferol and quercetin. The ethanol extracts of this species have shown central nervous system-depressant and hypotensive properties. This species has also been used as an anodyne and antispasmodic, diaphoretic, and diuretic expectorant and rubefacient and to promote healing of wounds. Externally it is used for removing warts and for rheumatic pain.

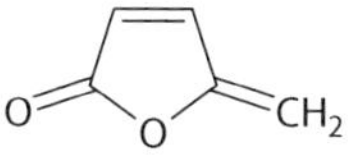

protoanemonin

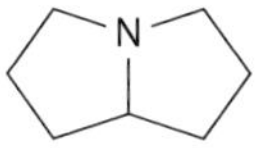

pyrrolizidine

MYRIOPHYLLUM SPICATUM LINNAEUS

Phylum: Magnoliophyta
Class: Magnoliopsida
Order: Haloragales
Family: Haloragaceae
Common name: Eurasian water milfoil
Distribution: Europe, Asia, North Africa
Habitat: Still or slow-moving waters
Description: This submersed species has slender stems of 3 m long. The leaves are also submersed (15–40 mm long) and are borne in whorls of 4. They are also pinnate with many leaflets, which are thread-like. The flowers (5–6 mm long), which are borne in leaf axils, are orange-red and monoecious.
Compounds and activities: A total of six quercetin glycosides, two kaempferol glycosides, a polyphenol, ellagic acid, gallic acid, pyrogallol, and (+)-catechin were isolated from this species. Of them, four quercetin glycosides, viz., quercetin 3-O-glucosyl-(1->2) xyloside, quercetin 3-O-glucosyl-(1->4)-rhamnoside, quercetin 3-O-rhamnoside (quercetin), and quercetin 3-O-xyloside, were found to show antioxidant properties. They also exhibited significant antimicrobial activities against certain Gram-negative and Gram-positive bacterial strains as well as fungi. Moreover, they exhibited significant cytotoxic activity against colon, gastric leukemia, and lung carcinoma cell lines (Afifi et al., 2009).

MYRIOPHYLLUM VERTICILLATUM LINNAEUS

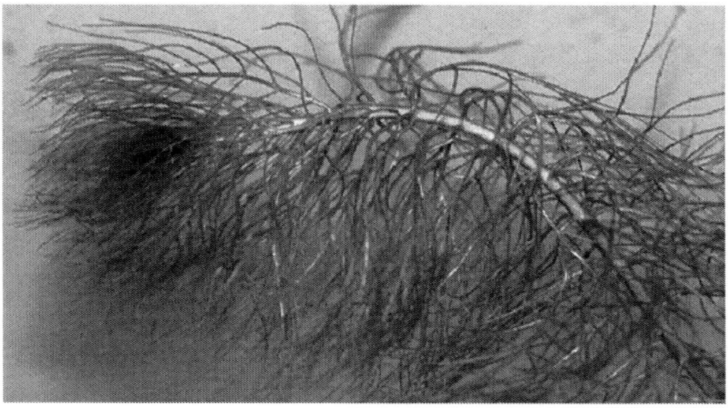

Phylum: Magnoliophyta Class: Magnoliopsida
Order: Haloragales Family: Haloragaceae
Common name: Whorled water milfoil, myriad leaf
Distribution: United States, United Kingdom, Asia, North Africa
Habitat: Semishallow ponds, lakes, marshes, ditches, slow-running streams.
Description: This perennial species grows to a maximum height of 1 m and possesses two types of leaves, viz., feathery, submerged and emergent. The flowers of this species are monoecious.
Compounds and activities: Bioactive phenolic compounds, viz., isomers of hexahydroxydiphenyl-di- and tri-galloylglucose, have been isolated from this species and possess antioxidative activities. The total phenolic compounds (TPCs) of apicals of this species range between 38 and 122 mg TPC/g dw (Bauer, 2011).

NYMPHAEA CANDIDA PRESL. & C. PRESL.

Phylum: Magnoliophyta Class: Magnoliopsida
Order: Nymphaeles Family: Nymphaeceae
Common name: Dwarf water lily
Distribution: Northern Europe, Southwest Asia, India, China, Siberia
Habitat: Quiet freshwater habitats, viz., ponds, lakes, and streams
Description: It is a perennial herb with erect, ascending, and unbranched rhizomes. The leaf blade (10–25 cm diameter) is suborbicular and papery. Floating flowers (10–20 cm diameter) are white and cup shaped with yellow stamens and crimson stigmas. Fruit is semiglobose shaped.
Compounds and activities: A total of 13 phenolic compounds, such as kaempferol, kaempferol 3-O-β-D-glucopyranoside, kaempferol 3-O-2″-O-galloylrutinoside, kaempferol 3-O-α-L-rhamnopyranoside, kaempferol-3-O-α-L-rhamnopyranosylglucopyranoside, kaempferol-7-O-β-D-glucopyranoside 3-O-α-L-rhamnopyranosylglucopyranoside, quercetin, quercetin-3-O-β-D-xylopyranoside, myricetin, and myricetin-3′-O-β-D-xylopyranoside, which possess antioxidative properties, have been isolated from this species. The ethyl acetate fraction of this species showed potent antioxidant capacity with IC_{50} values of 59.32, 24.48, and 86.85 μg/ml for O_2^-, OH, and H_2O_2 radicals, respectively (Zhao et al., 2011). The ethyl acetate fraction of this species has also shown hepatoprotective effects. This species has also been employed in traditional medicines for problems such as headaches, cough, hepatitis, and hypertension, as it possesses anaphrodisiac, anodyne, antiscrophulatic, astringent, cardiotonic, demulcent, and sedative characteristics. A decoction of the root is used in the treatment of catarrh and kidney pain. It can be used for sore throats and is also used as a poultice for boils. The flowers have a sedative effect on the nervous system and are used for treating insomnia and anxiety.

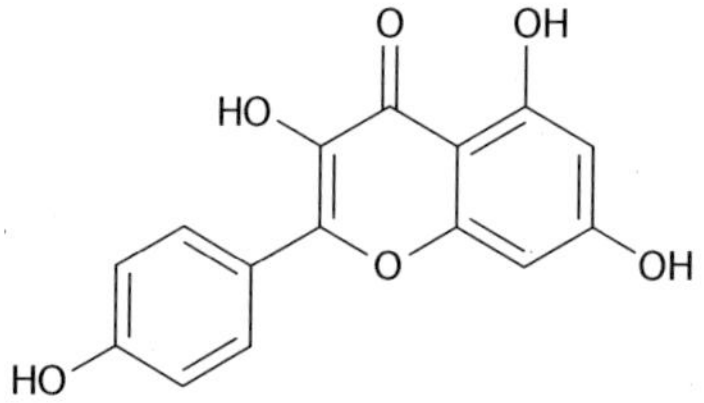

kaempferol

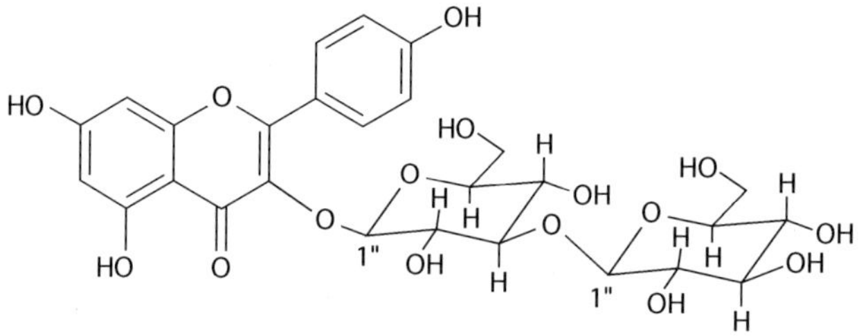

kaempferol 3-O-beta-D-glucopyranoside

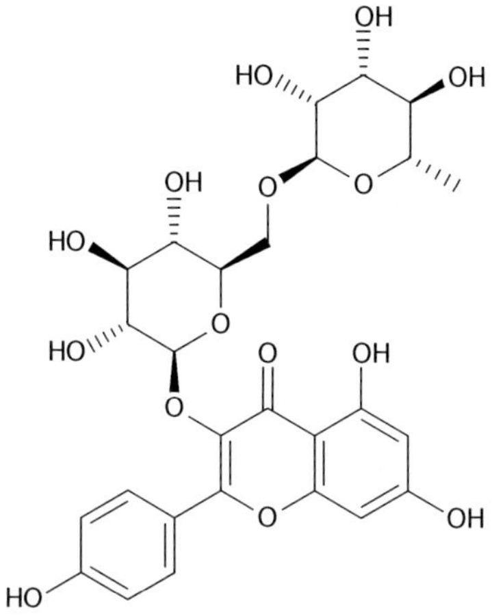

kaempferol 3-O-2"-O-galloylrutinoside

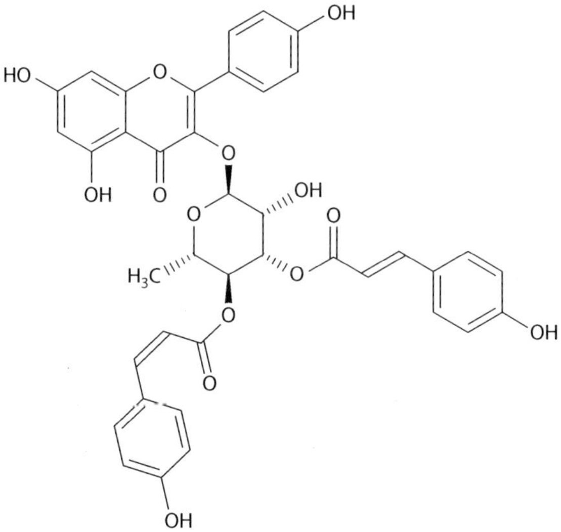

kaempferol 3-O-alpha-L-rhamnopyranoside

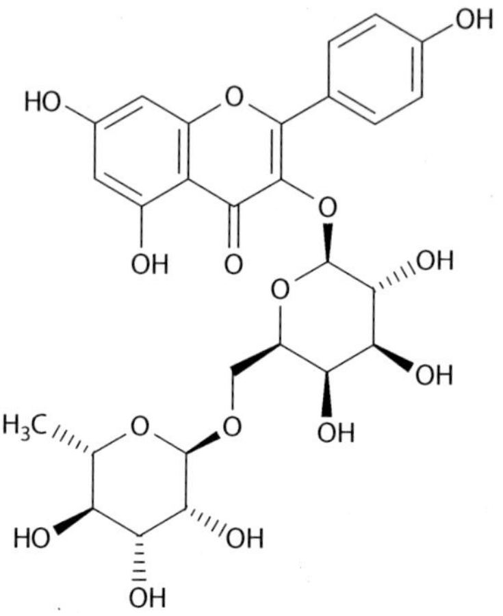

kaempferol-3-O-alpha-L-rhamnopyranosylglucopyranoside

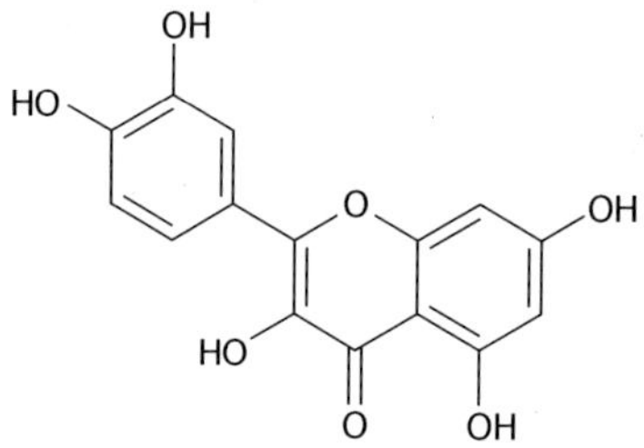

quercetin

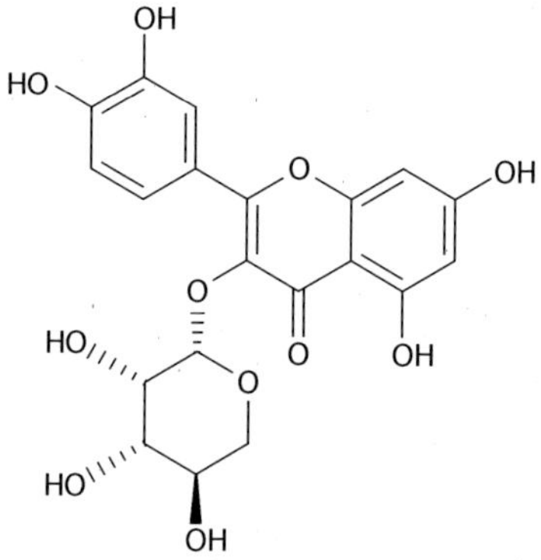

quercetin-3-O-beta-D-xylopyranoside

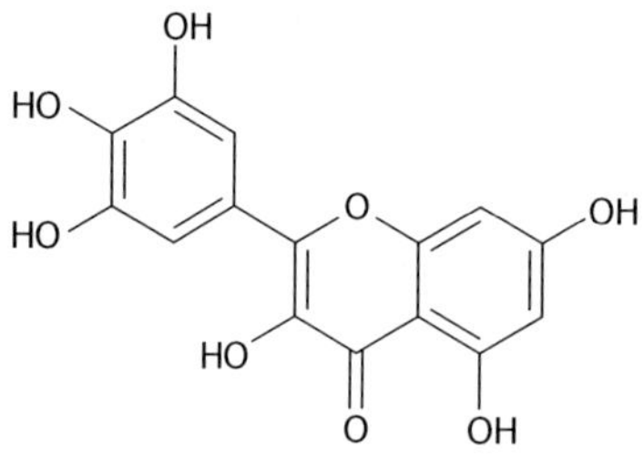

myricetin

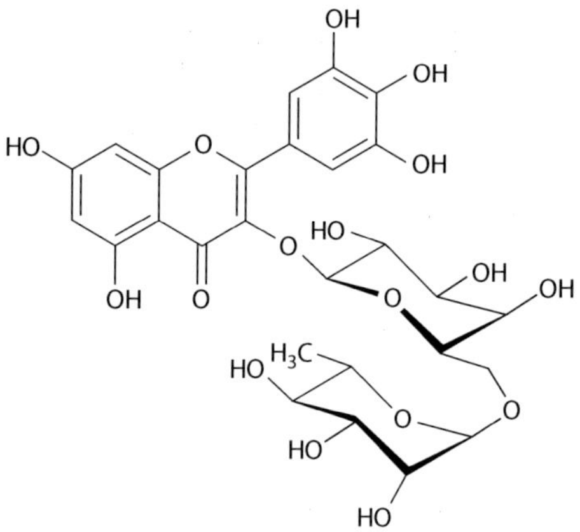

myricetin-3' -O-beta-D-xylopyranoside

NYMPHAEA COERULEA SAV

Phylum: Magnoliophyta
Class: Magnoliopsida
Order: Nymphaeles
Family: Nymphaeceae
Common name: Sacred blue lily, blue water lily
Distribution: East Africa, India, Thailand
Habitat: Freshwater habitats
Description: The leaves of this species measure 30–40 cm across and are broadly rounded. The flowers are 10–15 cm in diameter each.
Compounds and activities: The bioactive compounds, such as nuciferine (1,2-dimethoxy-aporphine), aporphine, apomorphine (6a-β-aporphine-10, 11-diol), and delphinidin 3-galloylgalactosides, possessing antioxidative properties have been isolated from this species (Fossen and Anderson, 1999).

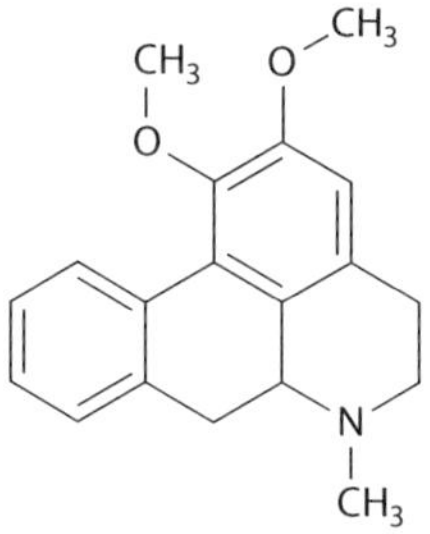

nuciferine (1,2-dimethoxy-aporphine)

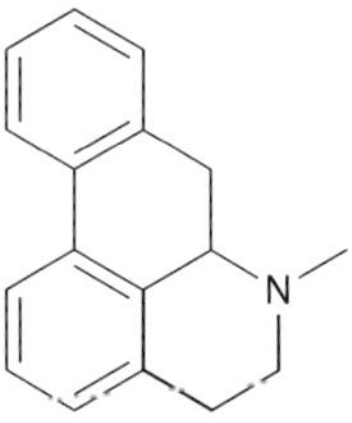

aporphine

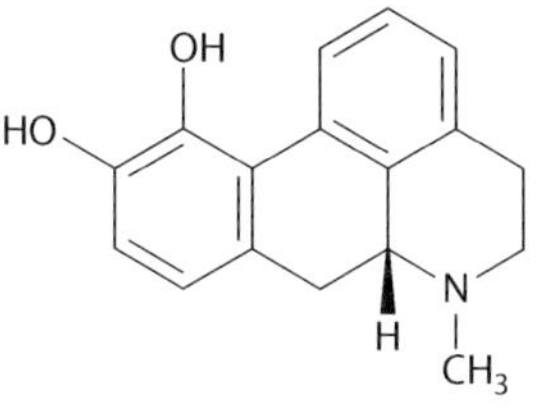

apomorphine (6a-beta aporphine-10,11-diol)

EURYALE FEROX SALISB.

Phylum: Magnoliophyta
Class: Magnoliopsida
Order: Nymphaeles
Family: Nymphaeceae
Common name: Fox nut
Distribution: Korea, Japan, Asia, Russia
Habitat: Wetlands, ponds, lakes
Description: This is a threatened perennial or annual species growing to a maximum height of 1 m. The leaves are very large (>1 m) and rounded. While their underside is purplish, their upper part is green in color. The bright and purple flowers are hermaphroditic.
Compounds and activities: Antioxidant compounds such as euryalins A–C, 5,7,4-trihydroxyflavanone, ferocerebrosides A [(2S,3R,4E,8E,2R)-1-O-(β-glucopyranosyl)-N-(2′-hydroxydocosanyl)-4,8-sphingadienine], B(2) [(2S,3R,4E,8E,2′R)-1-D-(β-glucopyranosyl)-N-(2′-hydroxytetracosanoyl)-4,8-sphingadienine], ferotectrimers C and D, and tocopherol trimers IVa and IVb have been isolated from this species (Song et al., 2011).

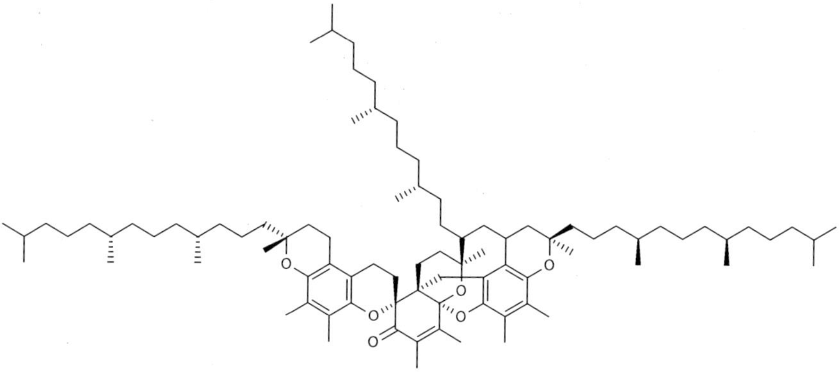

tocopherol trimers

IPOMOEA ASARIFOLIA (DESROUSSEAUX) ROMER & SCHULTES

Phylum: Magnoliophyta Class: Magnoliopsida
Order: Solanales Family: Convolvulaceae
Common name: Ginger-leaf morning-glory
Distribution: Africa, tropical Asia, Australia, South America
Habitat: Wet habitats
Description: The thick stem of this species is either prostrate or with twining tips. The leaves (7 × 10 cm) are circular or kidney shaped, and their base is cordate and the apex is broadly rounded. The inflorescence is axillary cymose, and its flowers (5–9 mm long) are elliptic and red-purple in color.
Compounds and activities: This species contains α-tocopherol, ascorbic acid, β-carotene phenolics, and ethers, which possess antioxidant properties. Bioactive compounds isolated include chanoclavine, chanoclavin-1-aldehyde, agroclavine, elymoclavine, and paspalic acid. The extracts of stem bark, leaves, and root bark of this species showed IC_{50} values of 50, 42, and 65 µl/3 ml, respectively. The average antioxidant activity expressed as ascorbic acid equivalent was found to be 1.7 mM (Atawodi and Onaolapo, 2010). The above compounds may also serve as acetylcholine esterase inhibitors. The roots of this species have also been reported to possess diuretic, emmenagogue, antidiabetic, and purgative properties (Meira et al., 2012).

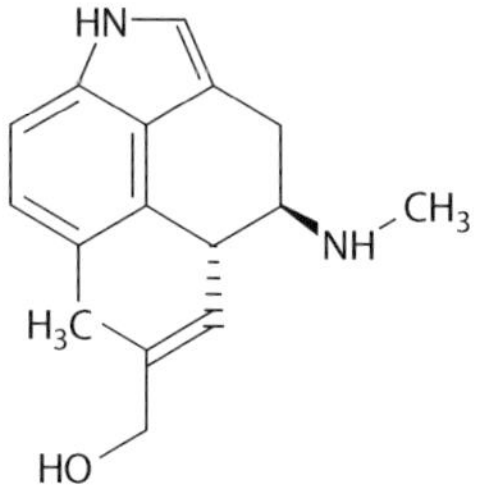

chanoclavine

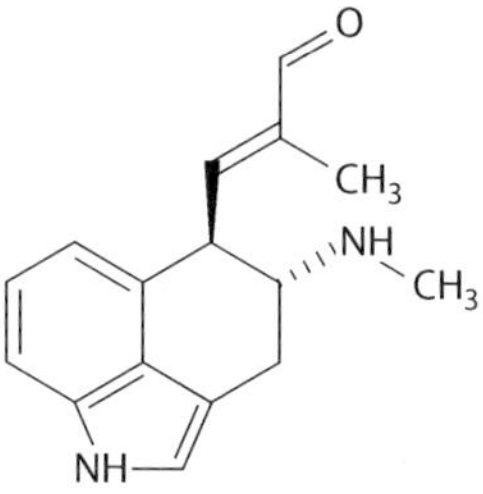

chanoclavin-1-aldehyde

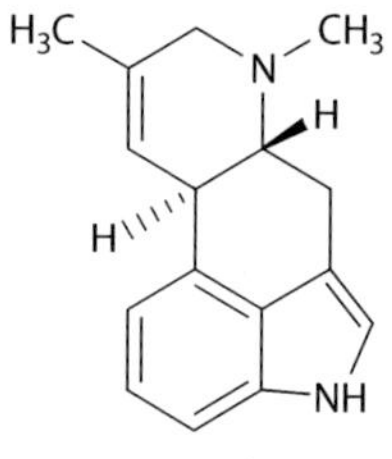

agroclavine

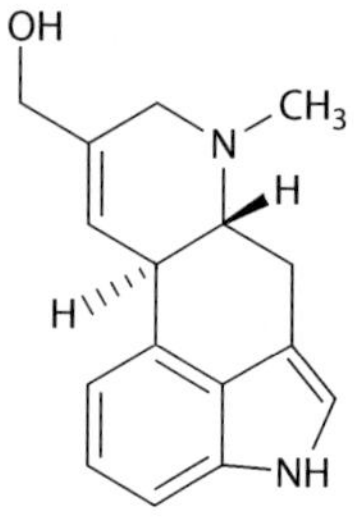

elymoclavine

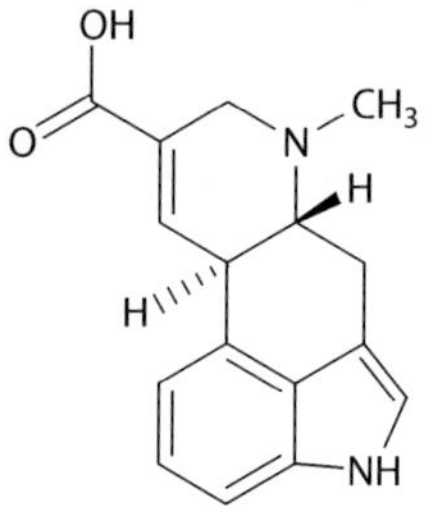

paspalic acid

MENTHA SPICATA LINNAEUS

Phylum: Magnoliophyta Class: Magnoliopsida
Order: Lamiales Family: Lamiaceae
Common name: Spearmint
Distribution: India, Europe, Southwest Asia, North America
Habitat: Stream banks, shorelines, wet, moist, and damp soils
Description: This perennial herb grows to 1 m. It possesses hairy or hairless stems and widespread, fleshy underground rhizomes. The leaves (9 × 3 cm) are broad with serrated margins. The flowers (2.5 × 3 mm), which are in slender spikes, are pink or white.
Compounds and activities: This species has yielded flavonoid compounds possessing antioxidant, anticancer, and chemopreventing activities. The bioactive compounds isolated from this species include kaempferol, naringenin, myricetin, luteolin, catechin, epicatechin, rutin, and apigenin. The antioxidant activity of SC-CUZ extracts (by the DPPH radical scavenging method) of the leaves of this species showed a value of 71.00 ± 2.65% at 200 bar pressure, 50°C, and 6 g/min co-solvent flow rate. However, all of the extracts showed significantly higher antiradical activities varying from

35.62 ± 0.34% to 72 ± 3.17% (Mandana et al., 2011). The tea prepared out of the extracts of this plant has antifungal properties besides treating hirsutism (excessive hairiness) in women. Further, this preparation has antiandrogenic properties, by reducing the level of free testosterone in blood, and digestive problems.

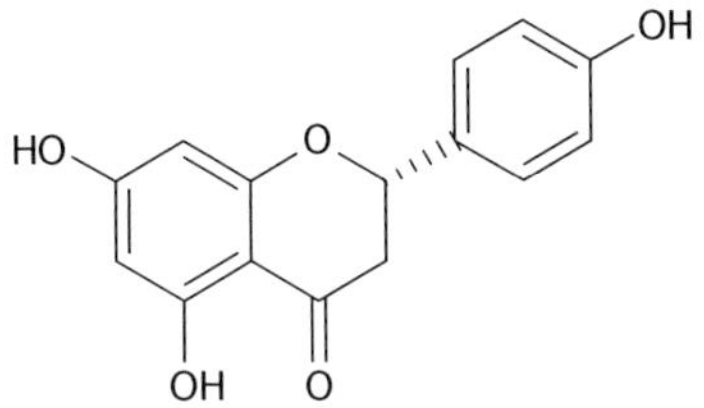

naringenin

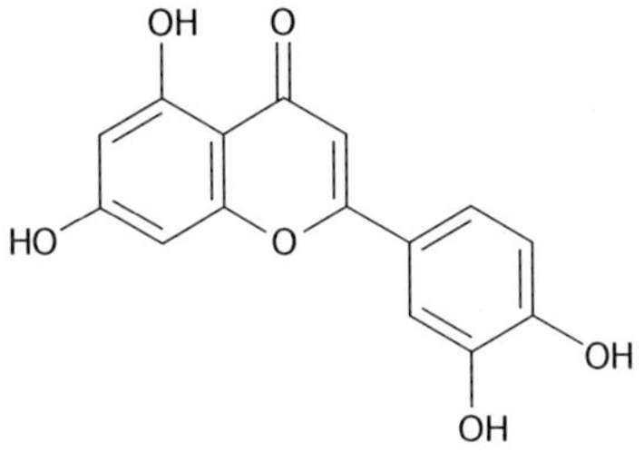

luteolin

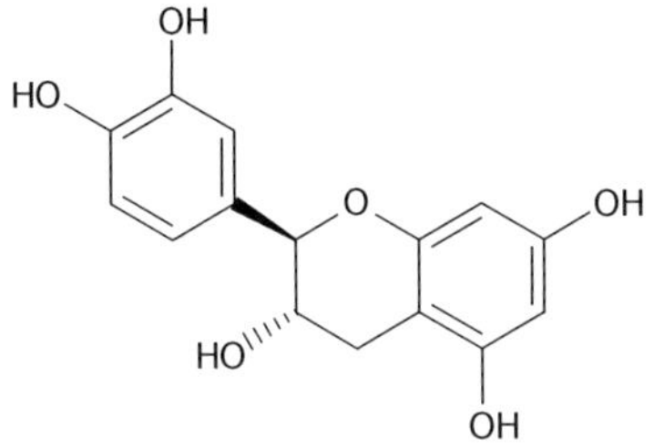

catechin

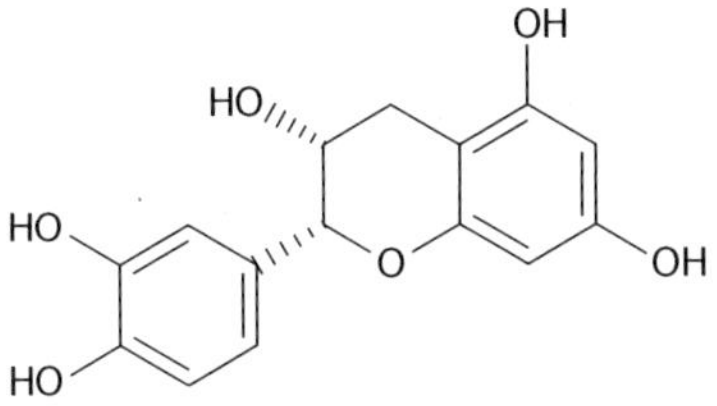

epicatechin

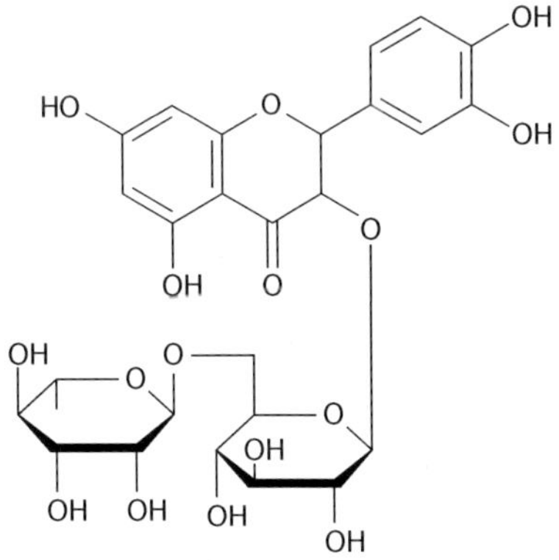

rutin

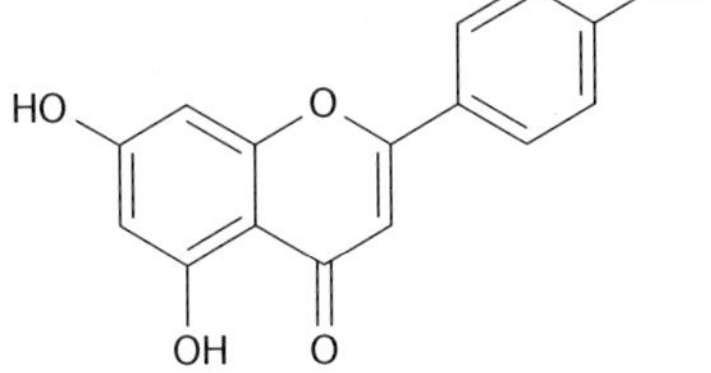

apigenin

SPHAERANTHUS AFRICANUS LINNAEUS

Phylum: Magnoliophyta Class: Magnoliopsida
Order: Asterales Family: Asteraceae
Common name: East Indian globe thistle
Distribution: Madagascar, Asia, tropical Africa
Habitat: Freshwater lakes, terrestrial habitats
Description: This species, which grows to a maximum height of 50 cm, has a stem that is robust, glabrous or pubescent curved, ascending, branched, and winged. Cauline leaves (4 × 2 cm) are ovate-oblong, and their bases narrowed and apices rounded. Marginal florets are numerous and corolla is tubular. There are three central florets.
Compounds and activities: The bioactive compounds isolated from this species include stigmasterol, chrysophenol D, 3,7-dimethoxy-4′5,6-trihydroxyflavone, chrysophenol C, 3α,5β-diangeloxoyloxy-7-hydroxy carvotanacetone, 1-angeloxoyloxy-3-[4′-angeloxoyloxy-3′-methoxy]-2-propene, and 1-angeloxoyloxy-3[4′-isopentanoloxoxoyloxy-3′-methoxy]-2-propene. Among them, compounds chrysophenol D, 3,7-dimethoxy-4′5,6-trihydroxyflavone, chrysophenol C, and 3α,5β-diangeloxoyloxy-7-hydroxy carvotanacetone showed free radical scavenging activity against the DPPH radical, and compounds chrysophenol D and 3α,5β-diangeloxoyloxy-7-hydroxy carvotanacetone exhibited cyclototoxicity against the KB cell (carcinoma cell line) (Kokpol et al., 2000).

SPHAERANTHUS INDICUS LINNAEUS

Phylum: Magnoliophyta Class: Magnoliopsida
Order: Asterales Family: Asteraceae
Common name: East Indian globe thistle
Distribution: India, Sri Lanka, Africa, Australia
Habitat: Marshes, rice fields, swamps
Description: It is a much branched and strongly scented annual species. The stem is with wings that are toothed. Leaves are ovate-oblong, narrowed at the base, dentate, and serrate. Flowers are in compound heads that are globose.
Compounds and activities: The aerial parts of this species showed bioactive compounds such as 5-hydroxy-7-methoxy-6-C-glycosylflavone, n-pentacosan, hentriacontane, n-triacontanol, β-sitosterol, stigmasterol, β′-D-glucoside of β-sitosterol, sphaeranthine, and a phenolic glucoside. The pharmacologic properties of this species include antioxidant, anxiolytic, analgesic, antipyretic, antihyperlipidemic, antimicrobial, hepatoprotective, immunomodulating, wound healing, psychotropic, bronchodilatory, antimacrofilaricidal, antiallergic, and antidiabetic. The free radical scavenging potential of the ethanolic extract of this species has been determined by *in vitro* models like ABTS, DPPH, superoxide dismutase, and nitric oxide radical activity. The calculated 50% inhibitory dose (ID_{50}) values were found to be 41.99, 33.27, 25.14, and 22.36%, respectively (Shirwalkar et al., 2006).

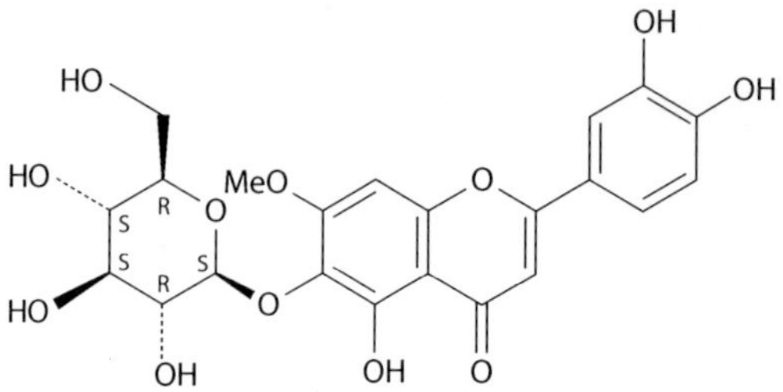

5-hydroxy-7-methoxy-6-C-glycosylflavone

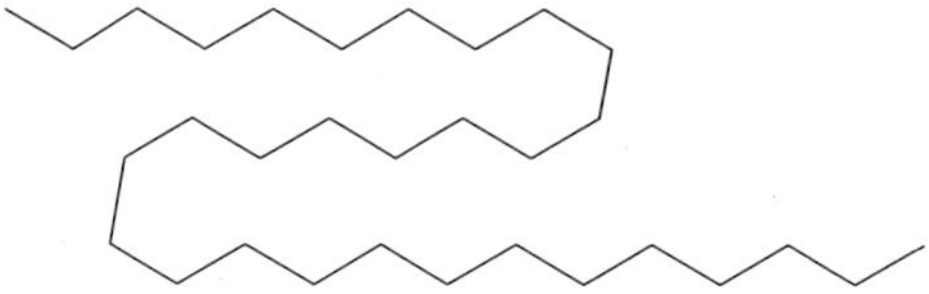

hentriacontane

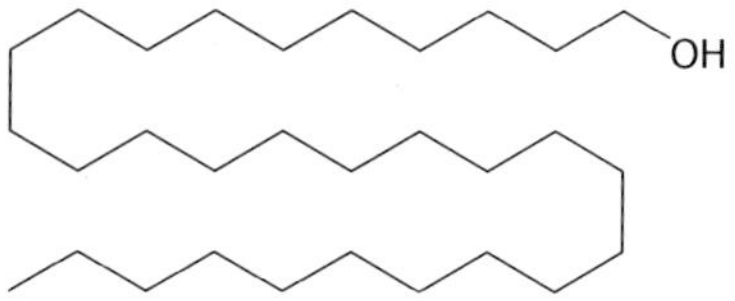

n-triacontanol

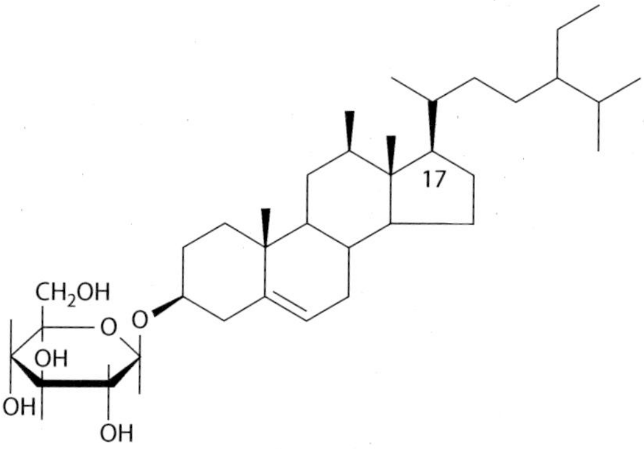

beta-D-glucoside of beta-sitosterol

GRANGEA MADERASPATANA (L) POIR (= *ARTEMISIA MADERASPATANA*)

Phylum: Magnoliophyta
Class: Magnoliopsida
Order: Asterales
Family: Asteraceae
Common name: Madras carpet
Distribution: Angola, Congo, China, Egypt, Ethiopia, Malaysia, Nigeria, Thailand
Habitat: Edges of ponds, canals, and ditches, dry river, pond beds
Description: It is a hairy branched herb, spreading from the roots, and growing to a maximum height of 70 cm. Leaves are alternate, deeply cut, stalkless, and divided into toothed lobes. Yellow flowering heads are borne opposite the leaves and are short stalked, rounded, and 10 mm across. Flowers are small and very numerous.
Compounds and activities: Two new 5-deoxyflavones, viz., 6-hydroxy-2′,4′,5′-trimethoxyflavone and 6-hydroxy-3′,4′,5′-trimethoxyflavone, and a known flavone, 7,2′,4′-trimethoxyflavone, possessing antioxidant, antifungal, analgesic, and antispasmodic activities, have been isolated (Patel et al., 2009; Rao et al., 2009). Its phenolic and flavonoid compounds are responsible for its free radical scavenging and antioxidant capacity. Using the methonolic extract of this species, IC_{50} values of 46.55 ± 1.67, 120.73 ± 0.694, 30.54 ± 1.11, and 209.73 ± 4.63 μg/ml were obtained for DPPH radical scavenging activity, NO scavenging activity, H_2O_2 scavenging activity, and β-carotene bleaching assay. The corresponding values with standard antioxidant, viz., ascorbic acid, were 24.96 ± 1.95, 236.37 ± 1.394, 57.34 ± 1.29, and 339.16 ± 5.30 μg/ml, respectively (Patel et al., 2009).

CAESULIA AXILLARIS ROXB

Phylum: Magnoliophyta
Class: Magnoliopsida
Order: Asterales
Family: Asteraceae
Common name: Pink node flower
Distribution: India, Pakistan
Habitat: Shallow waters, wet meadows, banks of irrigation canals, ditches, rice fields

Description: It is an erect annual herb, growing to a maximum size of 50 cm. The stem is stout pink and succulent. Leaves (15 cm long) are stalkless, alternately arranged, lance shaped, and pointed. Flowers are white and tubular with five linear petals.

Compounds and activities: The essential oil of this species exhibited free radical scavenging activity also recorded by 2,2-diphenyl-1-picrylhydrazyl assay, and its IC_{50} value was found to be 18 μl/ml (Mishra et al., 2012). Further, it showed a broad spectrum of fungi toxicity against fungi-causing postharvest deterioration of herbal raw materials. At 1.0 μl/ml, the oil showed complete inhibition of fungal growth, and aflatoxin B_1 production was inhibited at 0.8 μl/ml.

NELUMBO NUCIFERA GAERTN

Phylum: Magnoliophyta Class: Magnoliopsida
Order: Proteales Family: Nelumbonaceae
Common name: Sacred water lotus, Indian lotus
Distribution: India, China, Japan, Eastern Europe, Australia
Habitat: Ponds, rivers

Description: This perennial plant (maximum 1.5 m height) roots in soil at the pond and river bottom. The leaves, measuring 60 cm diameter, float on top of the water surface. The white and pink flowers (20 cm diameter) are hermaphroditic.

Compounds and activities: The ethanol root extracts of this species have yielded several bioactive compounds possessing antipyretic, anti-inflammatory, analgesic, antioxidant, antiproliferative, anticonvulsant, antiparkinsonian, hepatoprotective, hypoglycemic, and astringent properties. The highest levels of antioxidant (by 2,2-azino-bis(3-ethylbenzthiazoline-6-sulfonic acid) and 1,1-diphenyl-2-picrylhydrazyl assays) were noted in this species, and the values were 54.27 ± 6.1 and 21.98 ± 2.5 μm trolox equivalents/g dw, respectively (Park et al., 2009). Among the white and pink flowers, the former has shown high activity. The flower extracts of this species have yielded bioactive compounds belonging to flavonoids, alkaloids, tannins, phenols, saponins, phytosteroids, and glycosides. The bioactive compounds isolated from this species include nuciferin, nornuciferin, nelumborosides A and B, nelumstemine, dotriacontane, roemerin, liensinine, referine, lotusine, liriodenine, asimilobin, and pronuciferine. These compounds, apart from other activities, have also shown inhibitory actions on the secretion, aggregation, and suppression of platelets *in vitro*. The flavonoids of its leaf extract have been reported to attenuate body lipid accumulation and prevent obesity. The compound (S)-armepavine is a potential therapeutic agent for treating autoimmune crescentic glomerulonephritis (AcGN). In traditional medicines, its leaves are used for treating sunstroke, diarrhea, dysentery, fever, dizziness, and blood vomiting. Its fruit decoction is used for treating fever, heart problems, and for blackening hair. The plant extract is also used as an antidote for smallpox and for treating cholera, worm infection, and intermittent fever.

LINDERNIA ANAGOLLIS (BURN. F.) PENNELL

Phylum: Magnoliophyta Class: Magnoliopsida
Order: Scrophulariales Family: Scrophulariaceae
Common name: Yellow vein virus
Distribution: Continental and Southeast Asia, Northern Australia
Habitat: Along rivers, marshes, rice fields
Description: It is an annual herb that can grow to a maximum height of 40 cm. It is branched at the base. Branches are short, prostrate, or erect. Subsessile leaves are oblong-lance shaped to ovate with a maximum length of 50 mm. They are truncate at the base, with shallow dentate margins. Flowers are axillary and solitary, with one per leaf pair. The petal tube is 10 mm long. Its upper lip is entire and lower lip three lobed. The petal base is yellowish, and petal lobes are bluish purple or white.
Compounds and activities: The extract of this species exhibited strong scavenging DPPH radical activity with an IC_{50} value of 24 μg/ml (Shyur et al., 2005).

EPILOBIUM HIRSUTUM LINNAEUS

Phylum: Magnoliophyta
Class: Magnoliopsida
Order: Myrtales
Family: Onagraceae
Common name: Hairy willow herb
Distribution: Europe, North America, Asia
Habitat: Marshland, banks of rivers and streams
Description: It is a perennial plant growing to a maximum height of 2 m. The stems are robust, and branched with much hair. Leaves are also hairy (10 cm long and 4 cm wide) and have sharply toothed edges, but without stalk. Flowers (15 mm long) are large and purple-pink with four notched petals.
Compounds and activities: The values of its bioactive compounds, viz., caffeic acid, p-coumaric acid, and ferulic acids (phenolic acids), were found to be 23.1 ± 0.03, 38.3 ± 0.09, and 10.9 ± 0.29 mg/100 g dw, respectively. The values of quercetin and myricetin (flavonoids) of this species were found to be 214 ± 0.03 and 191 ± 0.24 mg/100 g dw, respectively. The proportional relation of flavonoid's content to that of phenolic acids in this species has been estimated as 82.5%. This species showed the highest scavenging of DPPH (antioxidant activity) due to its significant content of catechins and procyanidins (102 mg/100 g dw). The values of antioxidant activity of this species (total equivalent antioxidant capacity (TEAC)) (μm trolox/100 g dw) obtained by ABTS, DPPH, and FRAP were found to be 69.5 ± 1.22, 2021 ± 22.1, and 275 ± 1.11, respectively (Wojdylo et al., 2007).

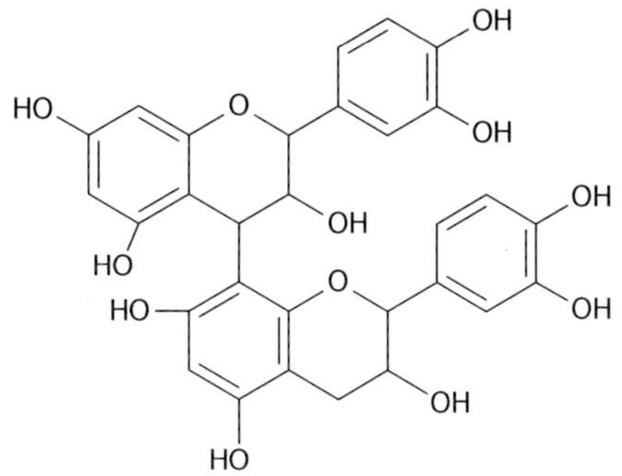

procyanidins

CENTELLA ASIATICA (L. URBAN)

Phylum: Magnoliophyta
Class: Magnoliopsida
Order: Apiales
Family: Machinlayaceae
Common name: Centella
Distribution: India, Sri Lanka, Australia, Indonesia, Iran, Malaysia
Habitat: Swamps
Description: It is a small herbaceous, annual plant. Stems are slender, green to reddish green, with creeping stolons. Leaves are green, and reniform with rounded apices. Flowers are pinkish to red in color, borne in small, rounded bunches (umbels) near the surface of soil. They are small and hermaphoditic with a size of 3 mm and with six corolla lobes.
Compounds and activities: The levels of phenolic and flavonoid compounds were 2.86 g/100 g and 0.361 g/100 g, respectively. Its aqueous extract (50 g/L) showed elevated DPPH scavenging activity with an IC_{50} value of 31.25 μg/ml. The aqueous extract showed a promising activity against mouse melanoma

($B_{16}F_1$), human breast cancer (MDA MB-231), and rat glioma (C_6) cell lines with IC_{50} values of 698, 648, and 1,000 μg/ml, respectively (Pittella et al., 2009). From the aerial parts of this species, a polyacetylene compound, viz., methyl 5-[(E)-9-hydroxy-1-(1-hydroxyhexyl)-2-methoxyundeca-3, 10-diene-5,7-diynyloxy] pentanoate (cadiyenol), possessing cytoxic property was isolated. This compound induced apoptosis (63%) independent of the cell cycle regimen in mouse lymphoma cells (P388D1) at 28 μM (IC_{50} = 24 ± 2 μM) in 24 h (Govindan et al., 2007). Further, its asiatic acid and asiaticoside have been reported to cure skin diseases and cancer. Furthermore, its phenolic compounds and flavonoids possess antioxidant properties (Pittella et al., 2009).

JUNCUS ACUTUS LINNAEUS

Phylum: Magnoliophyta Class: Liliopsida
Order: Poales Family: Juncaceae
Common name: Spiny rush, sharp rush
Distribution: North Africa, Western Asia, North Europe, Southeast Europe, Southwest Europe, Northern Mexico
Habitat: Freshwater wetland, saline and subsaline wetlands
Description: This perennial monocot species has short roots and robust rhizomes and grows to a maximum height of 1.5 m, and it has a pith-filled stem. The leaves arise from the base at different angles, giving a globe shape. The flowers are of 2–4 mm diameter and hermaphroditic.
Compounds and activities: The bioactive compounds, viz., 8,8′-bidehydrojuneusol (1,1′,6,6′-tetramethyl-5,5′-divenyl-8,8′-biphinanthrene-2,2′,7,7′-tetraol) (a phenanthrenoid dimer) and a related phenanthrene monomer, a phenolic chromone, and five flavonoid derivatives, have been isolated from this species. Among them, the phenanthrenoid dimer possesses antioxidant properties (Behery et al., 2012).

HYGRORYZA ARISTATA (RETZ) NEES EX WIGHT & ARN.

Phylum: Magnoliophyta
Class: Liliopsida
Order: Poales
Family: Poaceae
Common name: Water grass
Distribution: China, Sri Lanka, Taiwan, Southeast Asia, India
Habitat: Swamps, wet grounds
Description: It is a glabrous floating grass growing to a maximum height of only 4 cm. The stem is spongy with feathery whorled roots at the nodes. Leaves (8 cm long) are linear or ovate-oblong, and obtuse panicles are 5 cm long. Spikelets are very narrow and 20 mm long.
Compounds and activities: This species contains important anti-inflammatory and antioxidant compounds such as (7S,8R,7R,8S)-1 cariol A_2-9-O-β-xylopyranoside (lignin), hygarine (indole alkaloid), (–)-epigallocatechin-3-O-gallate, and epicatechin-3-O-gallate. Among them, the compounds below showed free radical scavenging activities and compound (7S,8R,7R,8S)-1 cariol A_2-9-O-β-xylopyranoside inhibited superoxide anion generation and elastase release by fMLP/CB-induced human neutrophils with IC_{50} values of 19.33 ± 0.86 and 24.14 ± 1.59 μM, respectively (Chung et al., 2011).

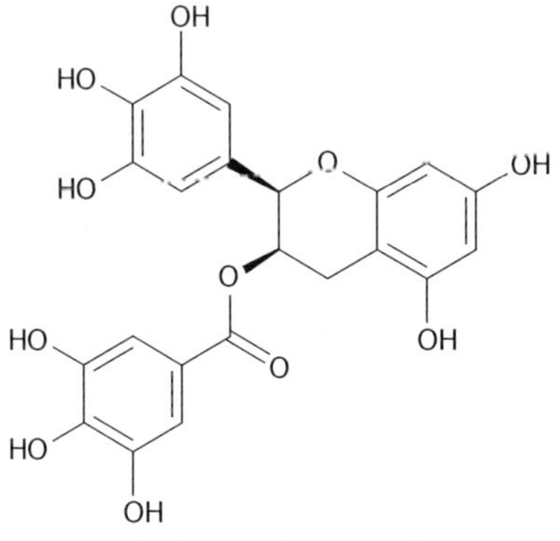

epigallocatechin-3-O-gallate

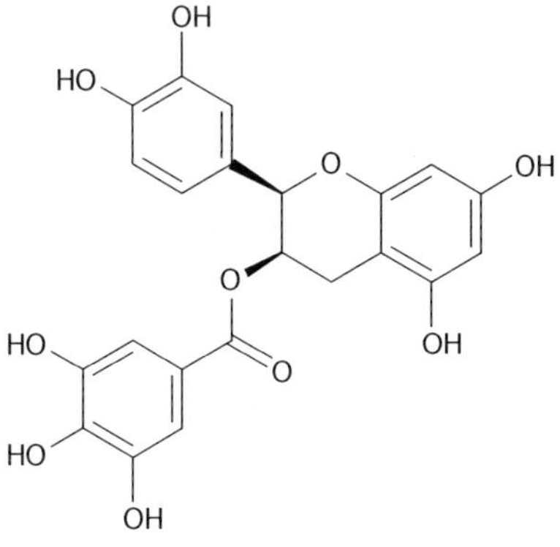

epicatechin-3-O-gallate

APONOGETON UNDULATUS ROXB.

Phylum: Magnoliophyta Class: Liliopsida
Order: Alismatales Family: Apogetanaceae
Common name: Undulate aponogeton
Distribution: Bangladesh, India, Sri Lanka, Burma, Thailand, Malaysia, Indonesia
Habitat: Still and slow-flowing shallow water bodies, flooded rice fields
Description: The rhizome of this submerged and rooted species is ovoid and measures 5 cm in length and 3 cm in breadth. The stem is single and erect, and its leaves (10–15 cm long), which have undulating margins, are very pale green. The flowers are white.
Compounds and activities: The crude extracts of this species showed antioxidant (free radical scavenging) and thrombolytic activities, besides possessing broad-spectrum antibacterial activities against *Bacillus cereus* and *Escherichia coli*. In a DPPH free radical scavenging assay, the extract exhibited potent antioxidant activity with an IC_{50} value 2.43 ± 1.06 μg/ml, while in ascorbic acid, the value was 2.14 ± 0.11 μg/ml. In thrombolytic activity (using the *in vitro* clot lysis assay method), the crude methanol extract was found to have significant ($p < 0.001$) thrombolytic activity at a dose of 10 mg/ml with a minimum effect of 20.23 ± 1.56%. The highest antibacterial effect was shown against *B. cereus* (zone of inhibition 12 ± 0.65 mm) followed by *E. coli* (zone of inhibition 10 ± 0.71 mm) (Chowdhury et al., 2011).

CAREX FOLLICULATA LINNAEUS

Phylum: Magnoliophyta Class: Liliopsida
Order: Cyperales Family: Cyperaceae
Common name: Northern long sedge
Distribution: Newfoundland, Quebec, Florida
Habitat: Meadows, swamps, edges of ponds
Description: This perennial species grows to more than 1 m. The leaves are 5–15 mm wide and are yellowish green to light green. While the terminal spike is staminate and peduncled, the pistillate spike is erect and subglobose.
Compounds and activities: The compounds isolated from this species include pallidol, kobophenol A, isoorientin, luteolin, quercetin, 3-O-methyl quercetin, and rutin. They possess antioxidant antibacterial and cytotoxic properties. The antioxidant activities of flavonoid compounds isoorientin, luteolin, quercetin, 3-O-methyl quercetin, and rutin showed IC_{50} values ranging from 50 to 200 μm, which were comparable to that of ascorbic acid (IC_{50} = 60 μm) and superior to those of the resveratrol derivatives (compounds pallidol and kobophenol A) (IC_{50} = >1,000 μm) and butylated hydroxyl toluene (IC_{50} = 1,500 μm), a commercial antioxidant. In the cytotoxicity and antibacterial assays compound luteolin (IC_{50} for human colon, 116 = 45 μm) and that shown in the structure of chanoclavine on page 159 (IC_{50} for methicillin-resistant *Staphylococcus aureus* = 6.4 μm) were the most active, respectively. Considering these, the bioactive compounds of this species may be useful for nutraceutical purposes (Li et al., 2009).

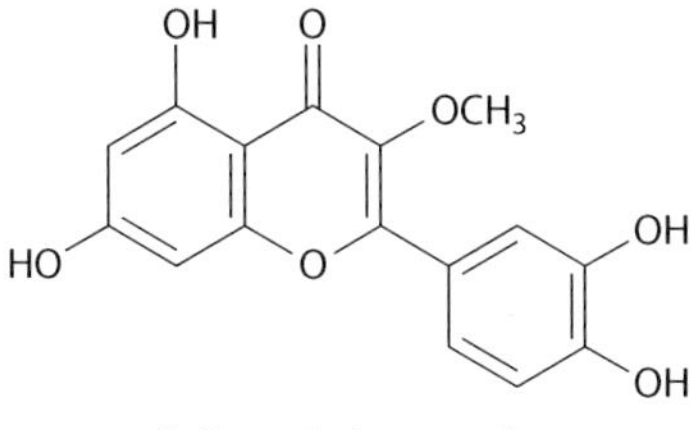

3-O-methyl quercetin

8 Anti-Inflammatory and Analgesic Compounds

Freshwater plants are of great use for therapeutic interventions in several inflammatory diseases. Inflammation is the response of the organism to invasion by foreign bodies such as bacteria, viruses, and parasites. Inflammatory response is a protective reaction to irritation, injury, or infection, which are characterized by redness, heat, swelling, pain, and loss of function. Inflammatory mediators of freshwater plants include arachidonic acid metabolites, peptides, cytokines, excitatory amino acids, etc.

APHANIZOMENON FLOS-AQUAE RALFS EX BORNET & FLAHAULT

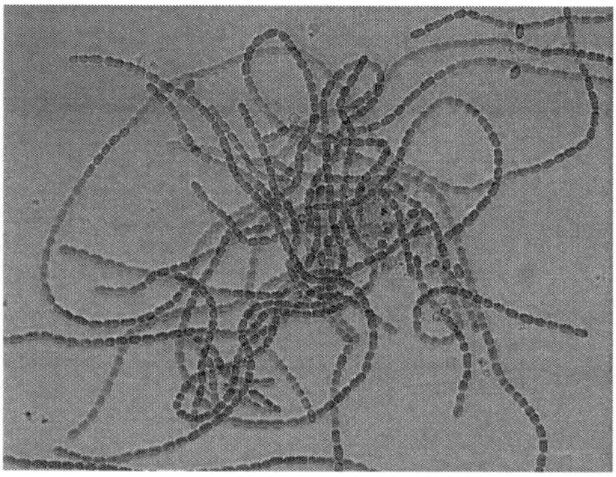

Phylum: Cyanobacteria Class: Cyanophyceae
Order: Nostocales Family: Nostocacea
Distribution: Europe, North America, Southwest Asia, Asia, Australia, New Zealand
Habitat: Freshwater systems
Description: This edible species is represented by both toxic and nontoxic forms. The organisms are microscopic, unicellular, and photosynthetic.
Compounds and activities: This species contains essential fatty acids, active enzymes, vitamins, amino acids, minerals, proteins, complex carbohydrates, and phytochemicals, which are known for their various activities, including antidepressant activity. The omega-3-alphalinolenic acid of this species possesses anti-inflammatory activity to support the immune system and to improve mental clarity and mental energy, and inhibits formation of inflammatory prostaglandins and arachidonate metabolites (Drapean, 2003). Further, its high concentration of phycocyanin inhibits the *in vitro*

growth of tumor cell lines and serves as an antioxidant (Singh et al., 2005). The crude extracts of this species containing chlorophylls stimulate liver function, reduce blood cholesterol level (hypocholesterolemic), inhibit activity of maltase and sucrose in the digestive tracts of test animals, and accelerate recovery from mild traumatic brain injury.

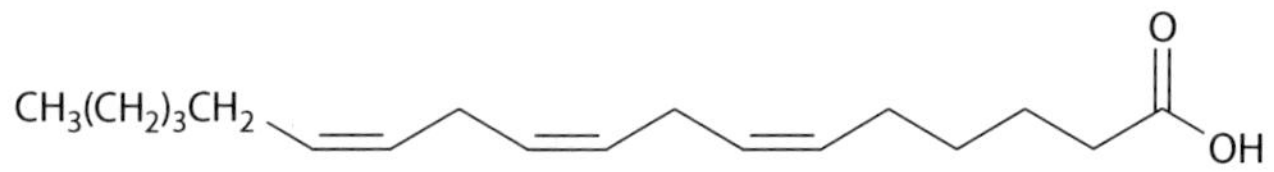

omega-3-alphalinolenic acid

IPOMOEA ALBA LINNAEUS

Phylum: Magnaliophyta
Class: Magnoliopsida
Order: Solanales
Family: Convolvulaceae
Common name: Moon flower
Distribution: Tropical America
Habitat: Margins of mangrove swamps, fresh- and saltwater marshes
Description: It is a fast-growing annual herbaceous vine. Leaves are ovate or circular in shape with a cordate base. Flowers are white and funnel shaped with a diameter of 15 cm.
Compounds and activities: The bioactive compounds isolated from this species include ipabine, ipalbidine, isoipomine, ipalbidinium, E-ipomine, Z-ipomine methoxy ipomine, dimethoxy ipomine, and ipohardine, which possess analgesic properties (Meira et al., 2012).

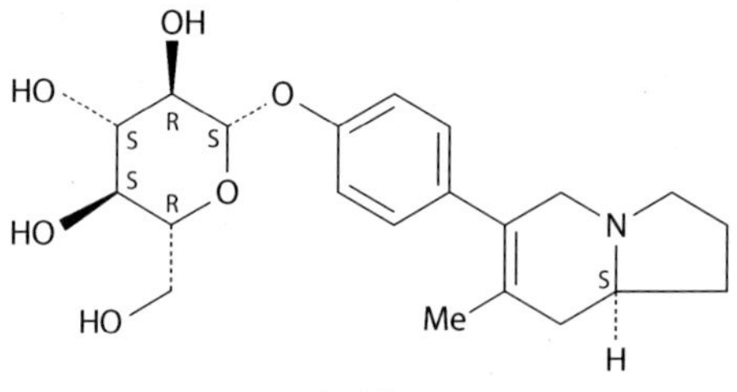

ipabine

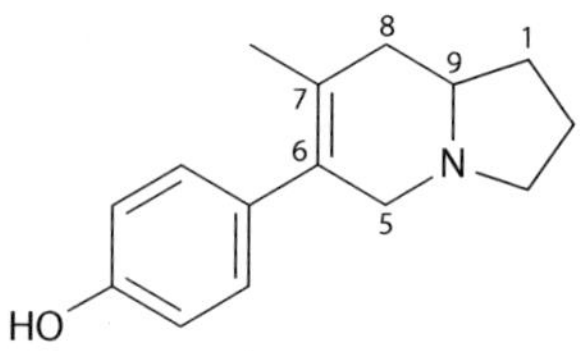

ipalbidine

IPOMOEA MURICATA (LINNAEUS) JACQUIN

Phylum: Magnaliophyta
Class: Magnoliopsida
Order: Solanales
Family: Convolvulaceae
Common name: Purple moon flower
Distribution: India
Habitat: Bank of streams and rivers
Description: It is a climber of 5 m long with purplish, slender twining stems. Leaves are 12 cm long and heart shaped. Purple flowers that are 5 lobed are in groups of one to few.
Compounds and activities: The bioactive compounds isolated from this species include ipalbine, ipalbidine, ipalbinium, and ipomine, which possess analgesic properties (Ysrael, 2003).

9 Antidiabetic Compounds

Diabetes, a major causative factor for death, may increase the risks of heart disease, kidney failure, blindness, and other serious health disorders. There are two types of diabetes mellitus (DM): type 1, juvenile diabetes mellitus (insulin-dependent diabetes mellitus (IDDM)), which is hereditary, and type 2, adult diabetes mellitus (non-insulin-dependent diabetes mellitus (NIDDM)), which occurs in older people. Many freshwater plants offer a variety of hypoglycemic (antihyperglycemic) compounds, such as alkaloids, glycosides, terpenes, flavonoids, etc., for combating both types of DM.

OSCILLATORIA PERORNATA SKUJA, 1949

Phylum: Cyanobacteria
Class: Cyanophyceae
Order: Oscillatoriales
Family: Oscillatoriaceae
Distribution: Australia, New Zealand
Habitat: Freshwater habitats
Description: The trichome of this species measures 4 and 10 μm in length and breadth, respectively.
Compounds and activities: From the pentane extracts of this species, volatile antidiabetic and antimicrobial compounds such as heptadecane (57%), 2-methylisoborneol (29.4%), benzaldehyde (1.2%), dimethyl disulfide (1%), dimethyl trisulfide (0.5%), and benzothiazole (0.6%) have been isolated (Tellez et al., 2001). Research is being conducted to find out their potential as pharmaceutical compounds.

heptadecane

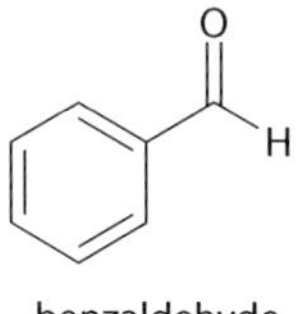

benzaldehyde

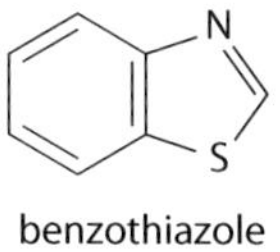

benzothiazole

LYNGBYA DENDROBIA BRUHL & BISWAS, 1923

Phylum: Cyanobacteria
Class: Cyanophyceae
Order: Oscillatoriales
Family: Oscillatoriaceae

Distribution: Australia, New Zealand

Habitat: Freshwater habitats

Description: In this filamentous species, the filaments are unbranched or slightly false-branched. They may be thick or rarely solitary. They are usually connected into thick-layered or leathery strata (mats). The trichomes (6–50 μm long) of this species are straight or slightly wavy.

Compounds and activities: The monounsaturated fatty acids, viz., palmitoleic and oleic acids, isolated from this species have been reported to prevent high glucose concentration and improve β cell secretory functions (Smith et al., 2008).

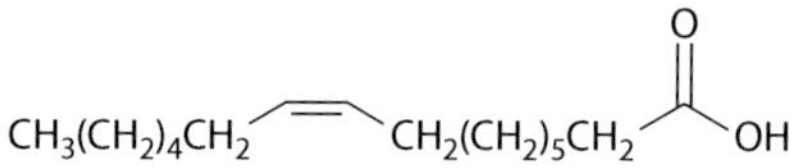

palmitoleic acid

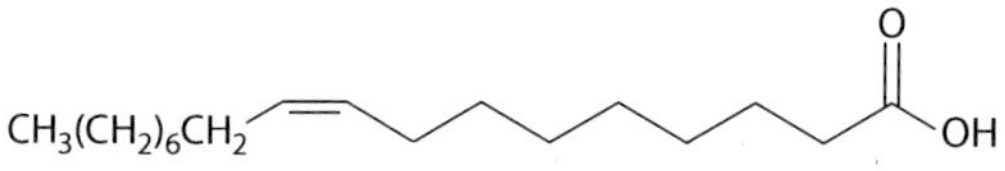

oleic acid

OSCILLATORIA CHLORINA KUTZING EX GOMONT, 1892

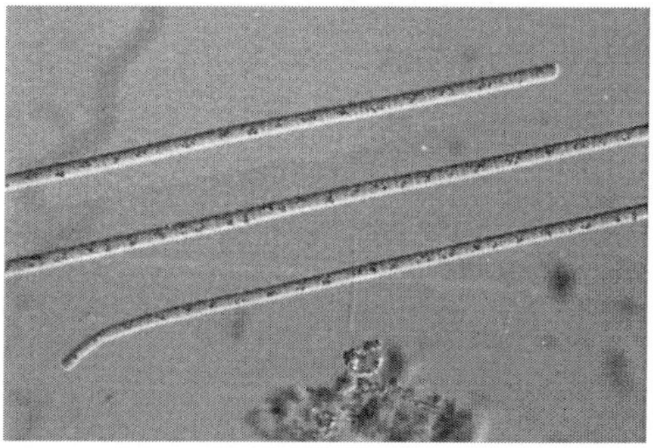

Phylum: Cyanobacteria

Class: Cyanophyceae

Order: Oscillatoriales

Family: Oscillatoriaceae

Distribution: Romania, Spain

Habitat: Freshwater systems

Description: The trichome (5 × 6 μm) of this species is straight or wavy and is not constricted at cell wall functions.

Compounds and activities: The bioactive compounds isolated from this species include oleic and linoleic acids, which possess antidiabetic activities.

OSCILLATORIA AMOENA (KUTZING) GOMONT, 1892

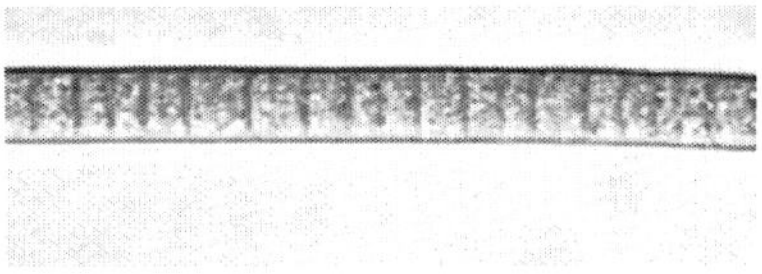

Phylum: Cyanobacteria
Class: Cyanophyceae
Order: Oscillatoriales
Family: Oscillatoriaceae
Distribution: Australia, New Zealand, Pacific Islands, Asia
Habitat: Freshwater habitats
Description: Each filament of this species consists of trichomes that are made of rows of cells. This filamentous species reproduces by fragmentation.
Compounds and activities: The bioactive compounds, viz., oleic acid and linoleic acid, isolated from this species have antidiabetic properties. Further, the extracts of this species possess antioxidant properties, besides serving as food additives and industrial chemicals (Smith et al., 2008).

OSCILLATORIA CALCUTTENSIS BISWAS, 1925

Phylum: Cyanobacteria
Class: Cyanophyceae
Order: Oscillatoriales
Family: Oscillatoriaceae
Distribution: India
Habitat: Freshwater habitats
Description: The thallus of this species is leathery brown and its trichomes (2 × 14 μm) are straight and parallel. The end cells are conical and pointed.
Compounds and activities: The bioactive compounds isolated from this species include palmitoleic acid, heptadecanoic acid, linoleic acid, and oleic acid. Among them, palmitoleic acid is known to improve the parameters of β cell formation and increase islet insulin content, as well as chronic and glucose-stimulated insulin secretions.

LYNGBYA LIMNETICA LEMMERMANN, 1898

Phylum: Cyanobacteria
Class: Cyanophyceae
Order: Oscillatoriales
Family: Oscillatoriaceae
Distribution: Europe, Southwest Asia, Australia, New Zealand
Habitat: Freshwater habitats
Description: This rock-mountain lake species is filamentous, and the filaments are singly, occasionally intertwined, irregularly coiled, or straight. The cells measure 8 μm in length and 2.5 μm width. The end cell is characteristically rounded and not tapering.
Compounds and activities: The bioactive compounds isolated from this species include oleic and linoleic acids. Of these, oleic acid is known to prevent high

glucose concentration and improve β cell secretory functions (Rajeshwari and Rajashekhar, 2011).

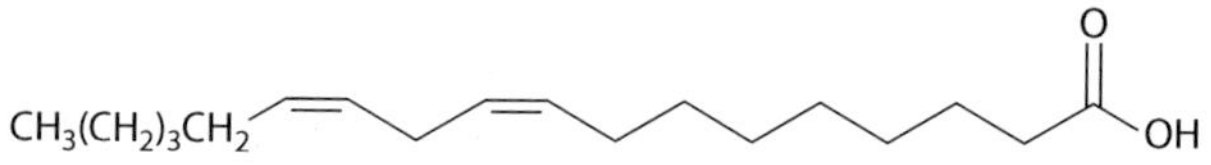

linoleic acid

PHORMIDIUM PURPURESCENS (KUTZING) GOMONT, 1890

Phylum: Cyanobacteria
Class: Cyanophyceae
Order: Oscillatoriales
Family: Phormidiaceae
Distribution: Romania, Spain
Habitat: Aeric or freshwater substrates
Description: The filaments of this species are unbranched and are solitary rarely. The cell contents make this species blue-green, brownish, pinkish, or violet.
Compounds and activities: The bioactive compounds isolated from this species include palmitoleic acid, oleic acid, and linoleic acid, which possess antidiabetic properties.

NYMPHAEA NOUCHALI BURM, F (= *NYMPHAEA STELLATA*)

Phylum: Magnoliophyta
Class: Magnoliopsida
Order: Nymphaeles
Family: Nymphaeceae
Common name: Blue water lily, star lotus
Distribution: India, Thailand, Myanmar
Habitat: Static or slow-flowing freshwater habitats
Description: This day-blooming, nonviviparous species possesses submerged stems and roots. The leaves (20–30 cm diameter) of this species may spread to a 3 m area, and they are rounded and green above with a dark underside. The flowers may be violet-blue, white, or purple in color.

Compounds and activities: The aqueous, methanol, and ethanol extracts of the dried and fresh flowers of this species have shown significant antioxidant activities. The values of H_2O_2 scavenging activity in terms of percentage of inhibition were found to be 54.73, 62.95, and 58.6% for dried flower extracts, respectively. Values for fresh flower extracts were 51.096, 61.428, and 58.47%, respectively (Mohan Muruga Raja et al., 2010). The bioactive compound nymphayol (25,26-dinorcholest-5-en-3β-ol) of the flower extracts of this species possesses antidiabetic, anti-inflammatory, and hepatoprotective activities. Other compounds, viz., flavonoids, anthocyanin, tannins, gallic acid, astragalin, quercetin, and kaempferol have also been isolated (Nagaveni and Rao, 2010). This plant has also been employed in traditional medicines for curing liver and urinary disorders, menorrhagia (prolonged menstrual period), and blennorrhagia (excessive discharge—gonorrhea), besides serving as an aphrodisiac.

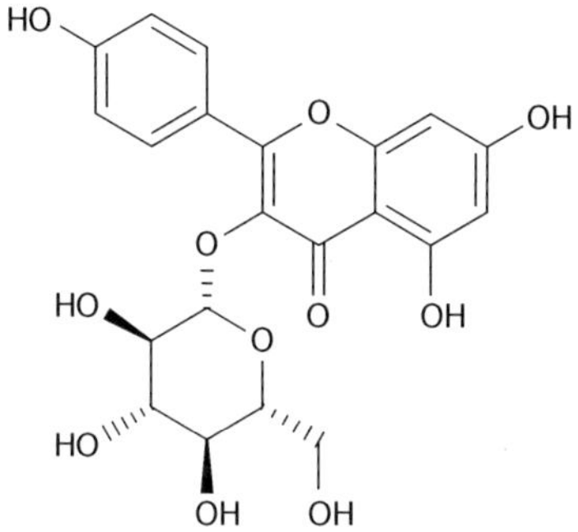

astragalin

10 Anticoagulant Compounds

Freshwater plants contain several bioactive compounds, such as calcium channel blockers, angiotension-converting enzyme inhibitors, angiotension receptor blockers, antihyperlipidemic agents, antiplatelet (anticlotting) compounds, and anticoagulant compounds, and these may be of great use in the production of drugs for cardiovascular diseases. Substances of drugs delaying the coagulation of blood are termed anticoagulants. The three broad categories of anticoagulants are calcium-sequestering agents (e.g., sodium citrate and sodium oxalate), prothrombopenic anticoagulants (e.g., sodium warfarin, coumarin, and their derivatives), and heparin and heparin substitutes produced from the phytoproduct of red seaweeds, viz., carrageenan.

MICROCYSTIS VIRIDIS (A. BRAUN) LEMMERMANN, 1903

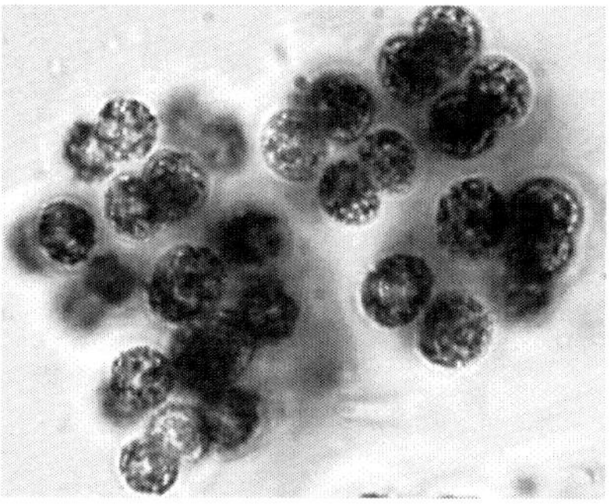

Phylum: Cyanobacteria Class: Cyanophyceae
Order: Chroococcales Family: Microcystaceae
Distribution: Europe, Southwest Asia, Australia, New Zealand
Habitat: Freshwater biotopes
Description: This colonial species forms harmful blooms. The cells (5–6 μm diameter) are spherical and without chloroplasts and pyrenoids.
Compounds and activities: The bioactive compounds isolated from this species include cyanoviridin RR, microcystin RR-YR-LR, microviridin, micropeptin 103, aeruginosin 102 A and B, and heptatoxic polypeptides. Among them, aeruginosin 102 A and B serve as thrombin inhibitors (anticoagulants) and micropeptin 103 as a chymotrypsin inhibitor (with IC_{50} of 1.0 μg/ml) (Namikoshi et al., 1992; Murakami et al., 1997).

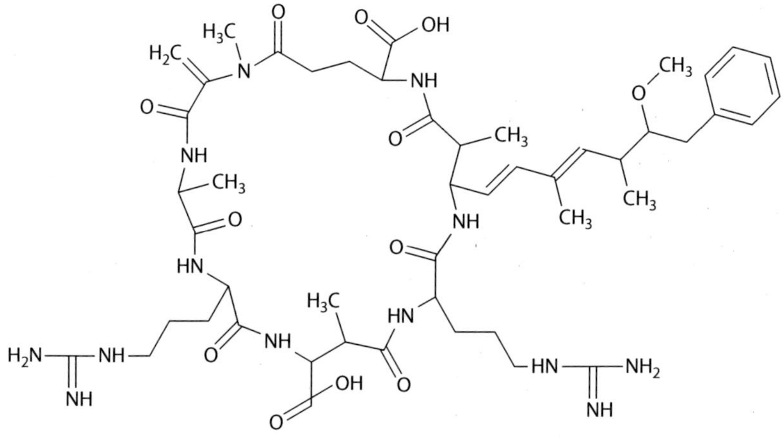

cyanoviridin RR

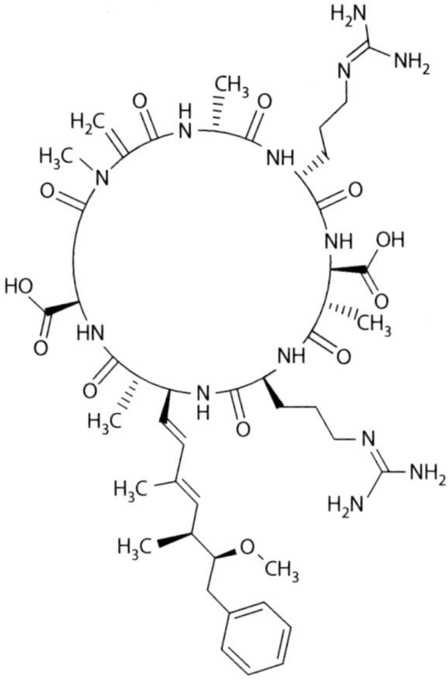

microcystin RR

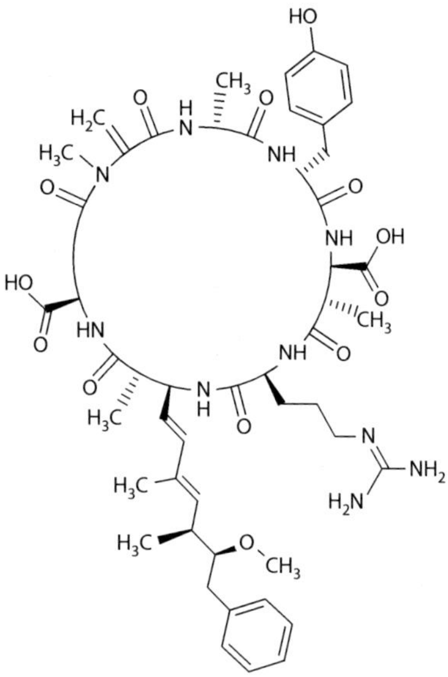

microcystin YR

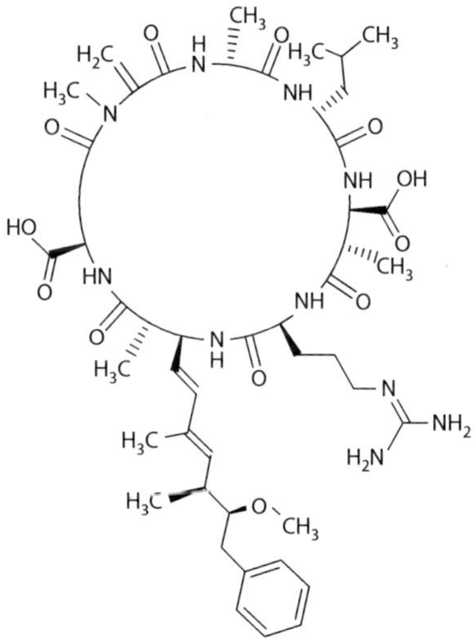

microcystin LR

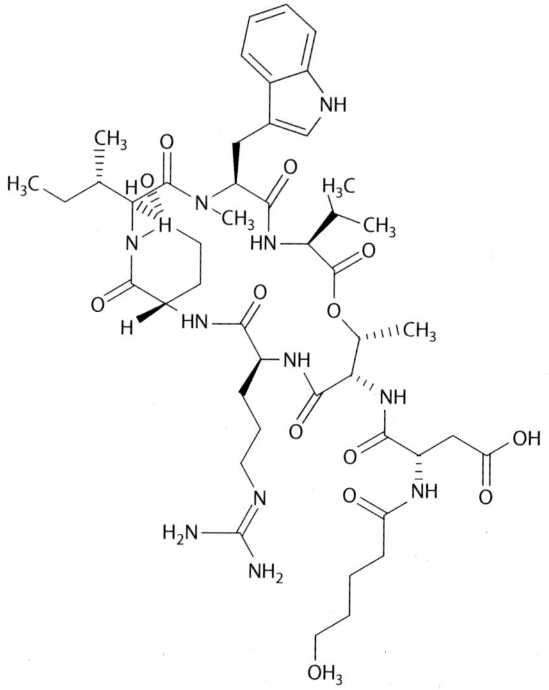

micropeptin

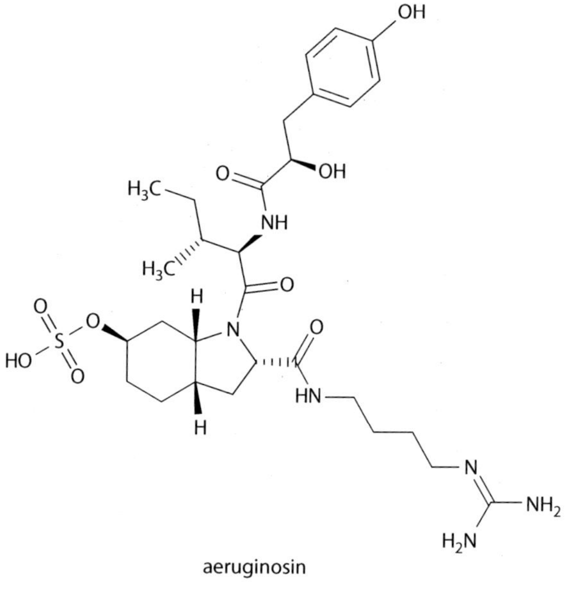

aeruginosin

NYMPHAEA AMPLA (SALISBURY) DE CANDOLLE

Phylum: Magnoliophyta
Class: Magnoliopsida
Order: Nymphaeles
Family: Nymphaeceae
Common name: Dream herb, white lotus, dot-leaf water lily
Distribution: Southern Mexico, Brazil
Habitat: Lakes, natural wells, ponds, slow-moving waters
Description: This species is characterized by the presence of a thick rhizome, long-stem cordate leaves, and immersed white flowers (18 cm diameter). The leaves (50 × 50 cm) are petiolate, glabrous, and their blades are ovate to orbiculate. The seeds are globose to ellipsoid in shape.
Compounds and activities: The bioactive compound isolated from this species is aporphine, which is closely related to opiate apomorphine. The extracts of this whole plant possess anticoagulant and aphrodisiac properties (Santos et al., 2012).

NUPHAR VARIEGATUM EUGELM. DURAND

Phylum: Magnoliophyta Class: Magnoliopsida
Order: Nymphaeles Family: Nymphaeceae
Common name: Variegated pond lily, bull-head pond lily
Distribution: Canada, Northern United States
Habitat: Ponds
Description: This perennial plant species has a spongy rhizome, and its submerged stems have simple, arrow-shaped, alternate leaves (15 × 5 cm) that float on the surface of water. The yellow and globe- or saucer-shaped flowers have six sepals each.
Compounds and activities: The extracts of this plant species possess biologically active tannins that have anticoagulant and antimicrobial activities. A significant increase of partial thromboblastin time (PTT) (more than 20 min) was observed in this species (Staba et al., 1972). The microbial species inhibited by this species include the mycobacterium *Smegmatis*, bacterial species *S. aureus*, and fungal species *C. albicans*.

MYRIOPHYLLUM EXALBESCENS FERNALD

Phylum: Magnoliophyta Class: Magnoliopsida
Order: Haloragales Family: Haloragaceae
Common name: Water milfoil
Distribution: Europe
Habitat: Shallow to deepwater lakes, ponds, ditches, marshes, slow-moving streams
Description: This submersed perennial plant has simple to freely branched stems (30–100 cm long). The leaves, which are 1–3 cm long, are pinnately dissected. The flowering spikes are red or reddish purple. While the upper male flowers are with petals, the lower female flowers are without petals. The fruits are olive in color and are subglobose.
Compounds and activities: The extracts of this plant species showed the presence of cyanidin glycosides, anthocyanins, flavonols, and flavonoid sulfate. Among them, anthocyanins, flavonols, and flavonoid sulfate possess anticoagulant properties. A significant increase of partial thromboblastin

time (PTT) (more than 20 min) was observed in this species (Staba et al., 1972).

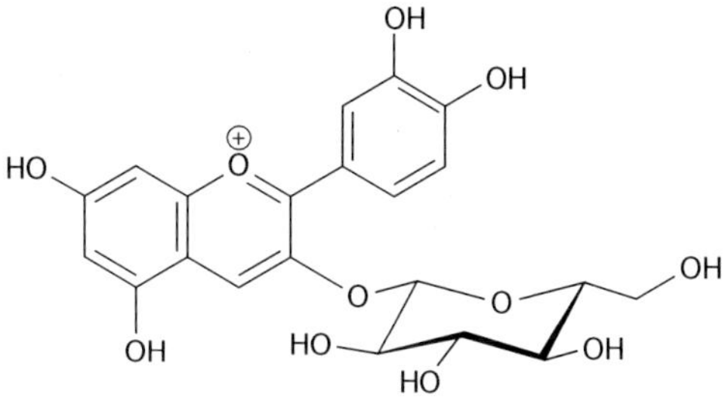

cyanidin glycosides

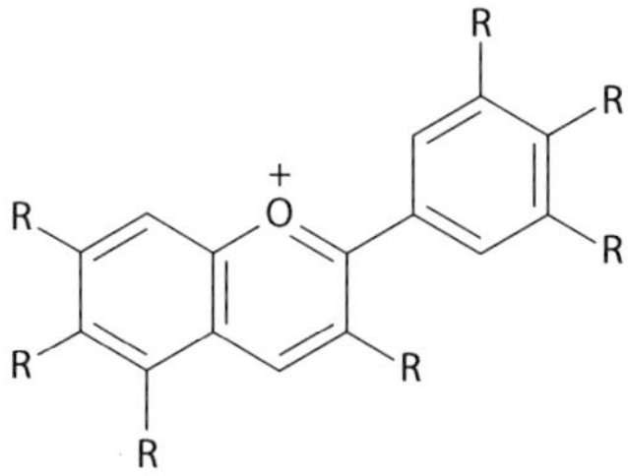

anthocyanins

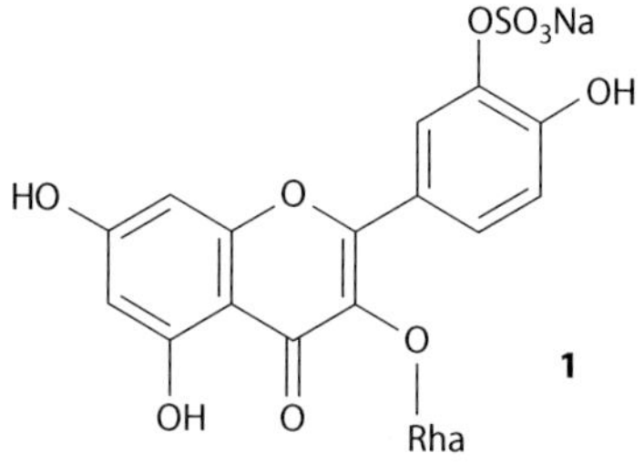

1

flavonoid sulfate

CAREX LACUSTRIS WILLDENON

Phylum: Magnoliophyta
Class: Liliopsida
Order: Cyperales
Family: Cyperaceae
Common name: Hairy sedge
Distribution: North America
Habitat: Shores of streams, ponds, lakes, open swamps

Description: It is a perennial species of colonial plant with long and creeping rhizomes. While the flowers and foliage are green, the fruits and seeds are brown. It grows to a maximum height of 2 m and has rhizomatous growth.

Compounds and activities: The extracts of this plant species have been found to possess anticoagulant properties. A significant increase of partial thromboblastin time (PTT) (more than 20 min) was observed in this species (Staba et al., 1972).

DAEMONOROPS ANGUSTIFOLIA (GRIFF) MART

Phyllum: Magnaliophyta Class: Liliopsida

Order: Arecales Family: Arecaceae

Common name: Water rattan

Distribution: Malay Peninsula, New Guinea, China, India, Thailand, Malaysia

Habitat: Swamps

Description: This climbing palm species has thickets of slender canes that are 40 m long. These canes possess many pinnate leaves with attractive narrow leaflets.

Compounds and activities: The bioactive compounds isolated from this species include 2-epi-lentiginosine, swainsonine, calystegines B1, B2, C1, and B3, and *N*-methyl-trans-4-hydroxy-L-proline, which possess anticoagulant properties (Gibbs et al., 1983).

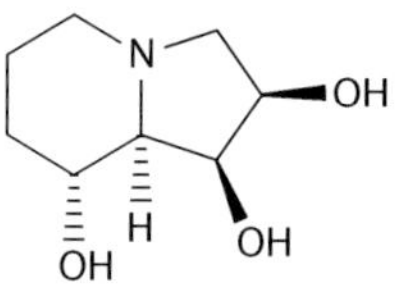

swainsonine

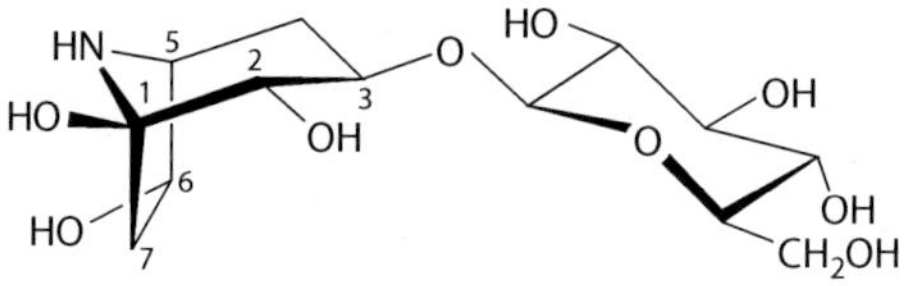

calystegines B1

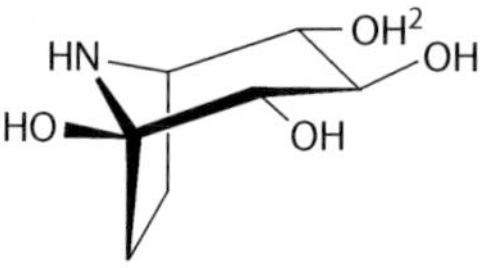

calystegines B2

11 CNS Disorders

The central nervous system (CNS)-related disorders include epilepsy, multiple sclerosis, spinal cord seizures, etc. Epilepsy, a chronic neurological disorder, is invariably associated with cerebral palsy (CP), which results from brain lesions. The different categories of phytopharmaceutical compounds relating to CNS disorders include reserpine, morphine, atropine, caffeine, physostigmine, benzodiazepine, etc.

MARSILEA MINUTA LINNAEUS

Phylum: Pteridophyta | Class: Filicopsida
Order: Hydropteridales | Family: Marsileaceae
Common name: Dwarf water cover, aquatic fern
Distribution: India, Africa, Trinidad, Tobago, Australia, New Caledonia
Habitat: Shallow pools, canals, edges of rivers, ditches, rice fields, brackish waters
Description: This perennial herb forms extensive but diffuse colonies. Leaves have erect terete petioles (15 cm long) and leaflets (2 × 2 cm).
Compounds and activities: This species contains bioactive compounds possessing antidepressant, antitumor, antioxidative, and anxiolytic activities. The extracts of this species prevent insomnia and other mental disorders. Marsiline isolated from this species possesses sedative and anticonvulsant properties. The extract of this species, when given orally once daily for 5 days at doses of 20 and 40 mg/kg, showed significant antidepressant activity in forced swim and learned helplessness models of depression (Jawaid et al., 2011). The extracts of this plant have also been reported to possess

hepatoprotective activity and are used in traditional medicines for the treatment of liver damage (hepatitis).

BACOPA MONNIERA L. PENNELL

Phylum: Magnoliophyta
Class: Magnoliopsida
Order: Lamiales
Family: Plantaginaceae
Common name: Water hyssop
Distribution: India, Nepal, Sri Lanka, China, Taiwan, Vietnam, Florida, Hawaii
Habitat: Wetlands, meadows flooded with water, damp and marshy places, shallow waters, brackish waters
Description: It is a small, creeping herb with small, oblong succulent and thick leaves with small light purple or white flowers of four to five petals each. This plant is propagated by cuttings.
Compounds and activities: The compounds isolated from this species are 3-O-stigmasterol-(6-O-palmitoyl)-β-D-glucopyranoside, bacopsides IX, I, and IV G, bacoparasaponins A, B, C, and D, and 1-octan-3-O-β-D-glucopyranoside, which possess antiepileptic, anxiolytic, antipyretic, analgesic, and antioxidative properties. It is an ayurvedic plant used for nervous and skin orders. The extracts of this plant are considered to be brain tonic, for epileptic and anxiety disorders, besides curing bronchitis, asthma, bowel syndrome gastric disorders, and improving respiratory and intellectual activities. Its antioxidative properties have been reported to protect certain types of cancers and cardiovascular diseases. Kaushik et al. (2009) have found that the alcoholic extract of this species could be a potent anticonvulsant drug.

IPOMOEA CARNEA JACE (= *IPOMOEA FISTULOSA*)

Phylum: Magnoliophyta Class: Magnoliopsida
Order: Solanales Family: Convolvulaceae
Common name: Pink morning glory
Distribution: North America, Brazil, Southern Asia, Bangladesh
Description: This dicot, perennial, vine shrub has dark green, heart-shaped leaves that measure 20 cm in length.
Compounds and activities: From this species, bioactive wound-healing flavonoids and polyhydroxylated alkaloids such as 2-epi-lentiginosine, swainsonine, calystegines B1, B2, B3, and C1, and *N*-methyl-trans-4-hydroxy-L-proline have been isolated. The compound marsilin isolated from this species showed antiepileptic (anticonvulsant) and sedative properties by protecting test animals from STR-induced convulsion (Navin and Rohatgi, 2005). The compound glycosidic saponin isolated from this species possesses anticarcinogenic and oxytoxic properties.

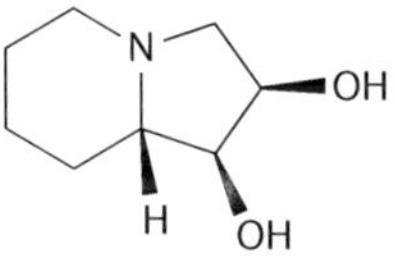

2-epi-lentiginosine

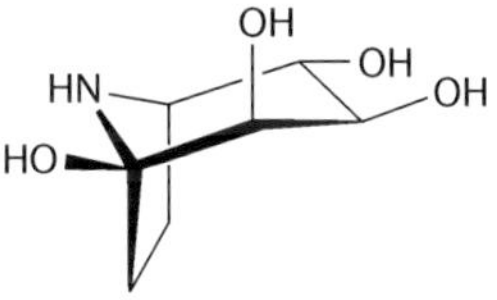

calystegines B3

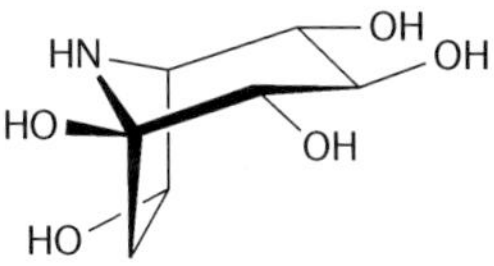

calystegines C1

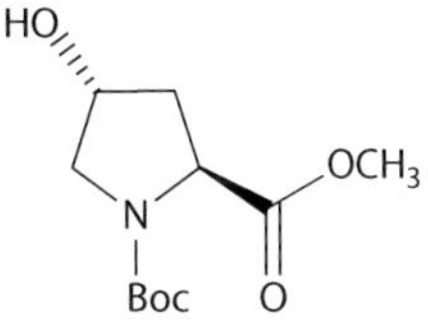

N-methyl-trans-4-hydroxy-L-proline

NYMPHOIDES MACROSPERMUM R.V. NAIR, 1965

Phylum: Magnoliophyta
Class: Magnoliopsida
Order: Solanales
Family: Menyanthaceae
Distribution: Endemic to Kerala (South India)
Habitat: Freshwater habitats
Description: This perennial plant has rhizomes and floating branches. The floating leaves are simple, orbicular, and cordate. The dioecious flowers are found aggregated on floating branches.
Compounds and activities: The alcohol extract of this species at doses of 500 and 750 mg/kg body wt significantly reduced the severity and increased the latency of convulsions induced by pentylenetetrazole (PTZ), and also reduced the time taken for recovery. The sedative effect of the plant extract was confirmed by a significant reduction in locomotor activity (Murali et al., 2007). The extracts of this plant species have also shown anti-helmintic activities.

LUDWIGIA REPENS J.R. FORST

Phylum: Magnoliophyta Class: Magnoliopsida
Order: Myrtales Family: Onagraceae
Common name: Red ludwigia, waterprimrose
Distribution: Central America, Northern America, Caribbean Islands
Habitat: Ditches, rivers, lakes, ponds
Description: It is an evergreen amphibious, polymorphic, mat-forming perennial herb with a creeping stem of 30 cm length. The leaves are oppositely arranged, and flowers are yellow with four petals each.
Compounds and activities: The methanolic extracts of this species have been reported to possess sedative and analgesic activities. Further, the extracts of this species may also have anxiolytic, sedative, or central nervous system-depressant activities. The methanolic leaf extract of this species decreased the locomotor activity of mice in hole cross, open-field, and elevated-plus maze tests. Further, the extract significantly minimized onset of sleep and maximized the duration of sleeping time when administered with thiopental sodium. The extract further produced a significant ($p < 0.05$) reduction of pain in all three models (Uddin et al., 2012).

NYMPHAEA ALBA LINNAEUS

Phylum: Magnoliophyta Class: Magnoliopsida
Order: Nymphaeles Family: Nymphaeceae
Common name: White water lily
Distribution: Europe, North Africa, Southwest Asia, India, China, Russia
Habitat: Large ponds, lakes
Description: This species grows in moderately deeper waters (0.5–1.5 m) and is rooted in mud. The circular floating leaves, which measure 30 cm in diameter, are dark green above and reddish below. The floating flowers are white.
Compounds and activities: The bioactive compounds isolated from this species include gallic acid, tannic acid, alkaloids, sterols, flavonoids, glycosides, hydrolyzable tannins, and high molecular weight polyphenolic compounds. The ethanolic extracts of this plant possess anxiolytic properties and are

used in the treatment of anxiety and insomnia disorders in humans. Further, these extracts have also shown *in vitro* antioxidant activity and skeletal muscle relaxant effects in mice. In traditional medicines, this plant is used as an aphrodisiac, anodyne, antiscrophulactic, astringent, cardiotonic, demulcent, and anti-inflammatory agent.

The ethanolic extract of this species significantly increased anxiolytic activity. In the foot shock-induced aggression test (FSIAT) the above extract was found to alternate aggressive behavior related to anxiolytic activity, such as number of vocalizations, leaps, rearing, biting/attacks, and facing each other in paired mice. Further, this extract also produced a skeletal muscle relaxant effect assessed by rota rod test (RRT) (Thippeswamy et al., 2011).

ECHINODORUS BERTEROI (SPRENGEL) FASSETT, 1955

Phylum: Magnoliophyta
Class: Liliopsida
Order: Alistamatales
Family: Alismataceae
Common name: Cellophane sword
Distribution: United States, Central America, West Indies
Habitat: Shallow and periodic waters, wet to moist swamps, ditches
Description: This monocot plant has an upright stem with both submersed and immersed leaves. The submersed leaves get modified into phyllodes, which are green, linear, or narrowly elliptical and are 50 cm long and 4 cm wide. The immersed leaves are light green, cordate, petiolated, and measure 12 cm (length) and 5 cm (width). The inflorescence is compound and has a white corolla measuring 1.5 cm in diameter.

Compounds and activities: The decoctions of the dried roots of this species have shown neuroleptic and antiepileptic properties. Repeated administration of 5% decoction (30 min intervals) significantly reduced the amplitude (muV) of the epileptic spikes induced by topical application of penicillin to the sensorimotor cortex, in curate-treated rats (Buznego and Perez-Saad, 2006).

JUNCUS EFFUSUS LINNAEUS

Phylum: Magnoliophyta Class: Liliopsida
Order: Poales Family: Juncaceae
Common name: Soft rush, common rush
Distribution: Most temperate countries
Habitat: Wet areas, marshes, streams, ditches
Description: This species grows in large clumps with a maximum height of 1.5 m. The stem is smooth and cylindrical, and the inflorescence is yellowish in color.
Compounds and activities: Several bioactive compounds, including phenanthrenes, have been isolated from the dry stem of this species. These compounds are dihydrodibenzoxepin, medullae juncus esters A and B, 5-α-spinasterol, β-sitosterol, effusol, ρ-coumaric acid, nobiletin, quercetin, and rutinose. Among them, effusol has shown anxiolytic and sedative properties (Liao et al. 2011; Wang et al., 2012). The above compounds are used as sedative and anxiolytic agents in Chinese medicines. Some of these compounds have also been reported to possess antitumor activities.

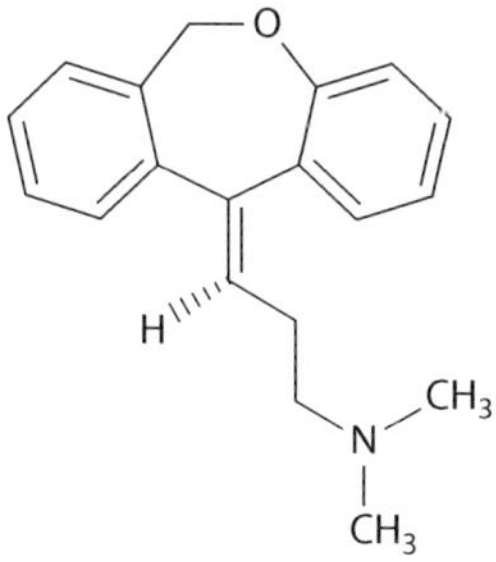

dihydrodibenzoxepin

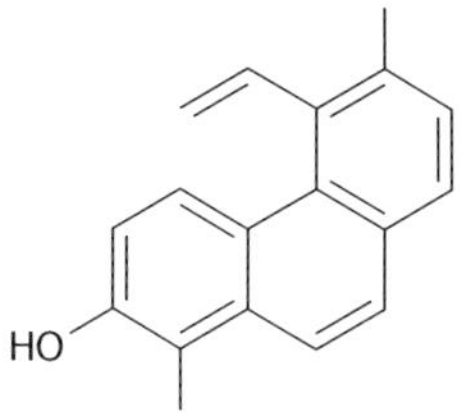

medullae juncus ester A, B

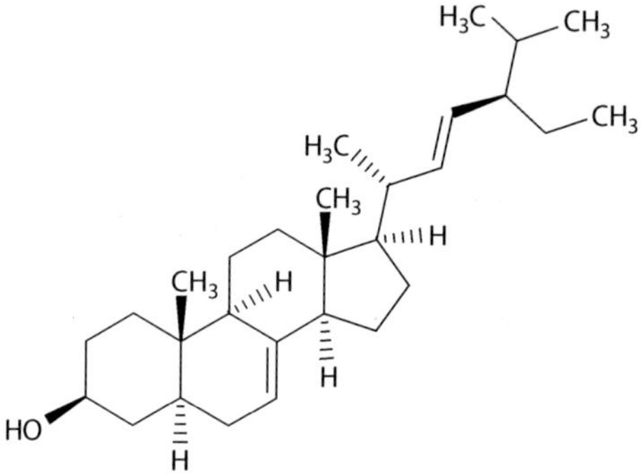

5-α-spinasterol

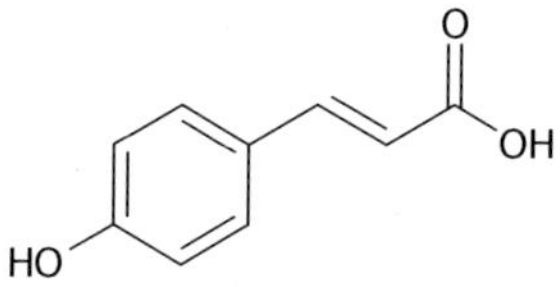

ρ-coumaric acid

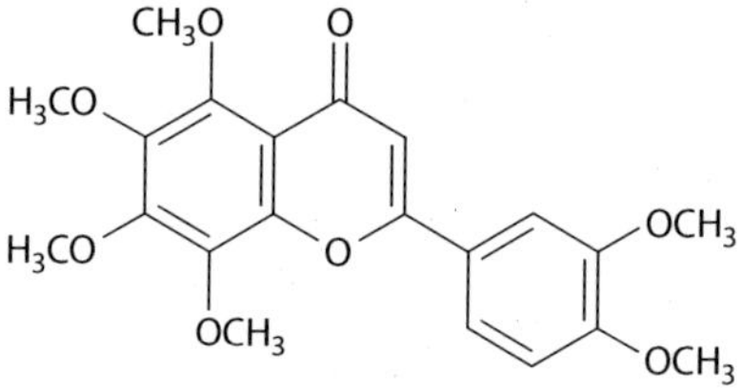

nobiletin

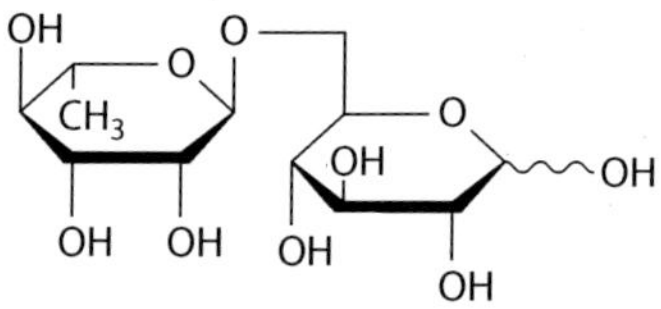

rutinose

12 Miscellaneous Compounds

Several freshwater plants also produce a number of compounds for the treatment of central nervous system (CNS)-related disorders. Such compounds include anti-epileptics (anticonvulsants), anxiolytics, antinociceptives, etc. Similarly, the above plant species produce bioactive compounds with antihyperlipidemic, antiplatelet, and hepatoprotective activities.

OSCILLATORIA AGARDHII GOMONT (= *PLANKTOTHRIX AGARDHII*)

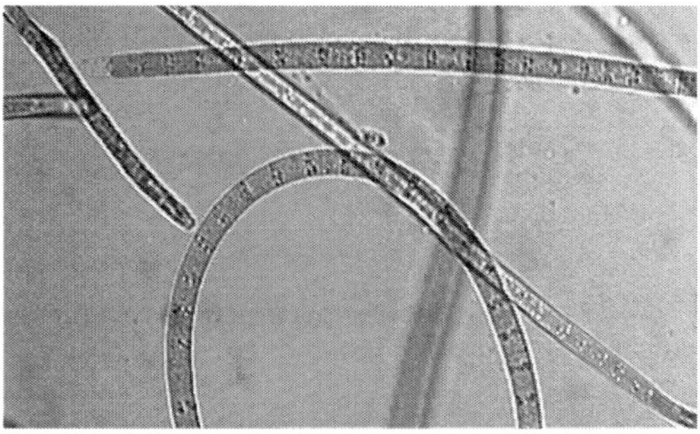

Phylum: Cyanobacteria Calass: Cyanophyceae
Order: Oscillatoriales Family: Oscillatoriaceae
Distribution: Europe, Southwest Asia, Australia, New Zealand
Habitat: Freshwater habitats, lakes
Description: The filaments (100 × 4 μm) of this species are solitary, straight or slightly waved, isopolar and free living, and without a sheath. The cells of trichomes (4 × 10 μm) are cylindrical and with slight tapering at the ends.
Compounds and activities: The bioactive compounds isolated from this species include oscillapeptide 97-A and -B, oscillacyclin, and anabaenopeptins B and F. Oscillapeptin of this species is a trypsin inhibitor (Sano et al., 1998).

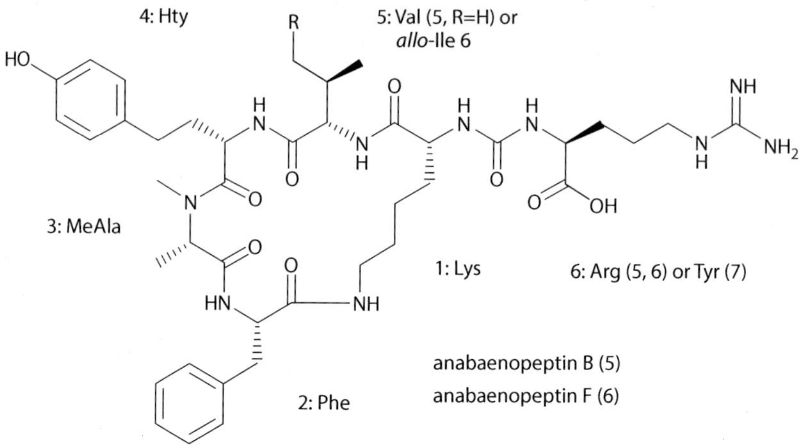

anabaenopeptins B, F

ANABAENA FLOS-AQUAE BREBISSON EX BORNET & FLAUHAULT, 1886

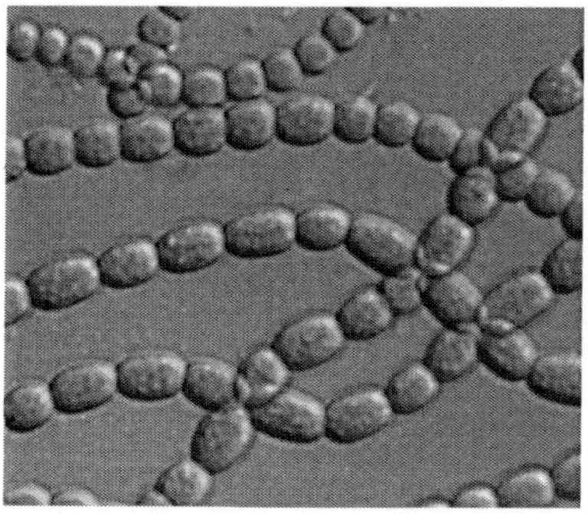

Phylum: Cyanobacteria Class: Cyanophyceae
Order: Nostocales Family: Nostocaceae
Distribution: Europe, South America, Africa, Southwest Asia, Central Asia, Southeast Asia, Australia, New Zealand
Habitat: Freshwater habitats, lakes
Description: This unicellular, microscopic species forms colonies that grow in filamentous clumps. The cells have no organized nuclei.
Compounds and activities: The bioactive compounds isolated from this species include anatoxin *a*, anabaenopeptins A and B, and microcystin-LR. Of them, anabaenopeptins A and B have been found to be protease (protein-phosphatase) inhibitors. The methanol extract of this species containing the compound 7-methyl heptadecane has shown antibacterial properties (Ostensvik et al., 1998). Anatoxin *a*, a guanidinemethyl phosphate ester, is an inhibitor of acetylcholinesterase.

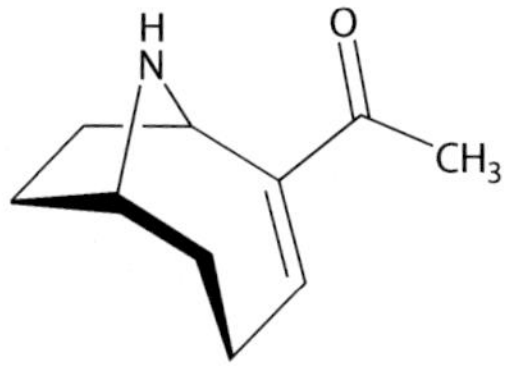

anatoxin-*a*

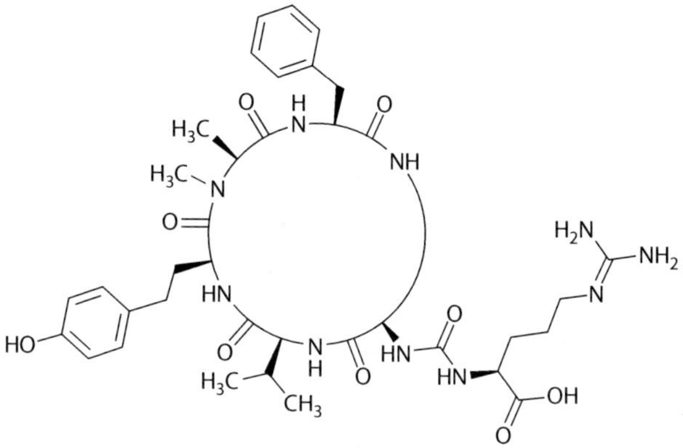

anabaenopeptin A

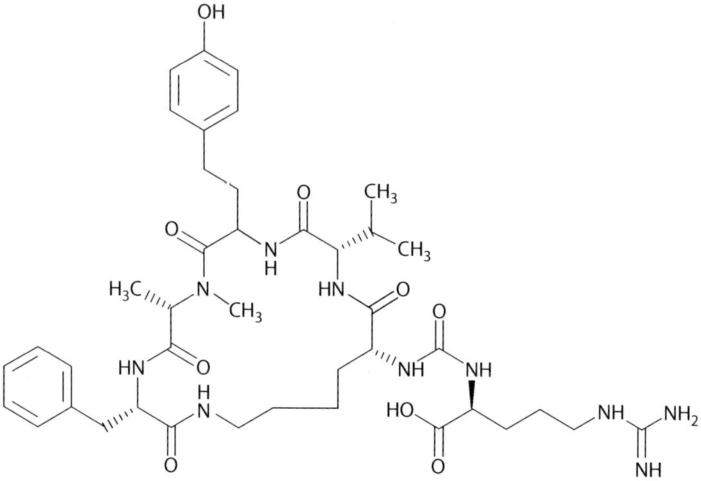

anabaenopeptin B

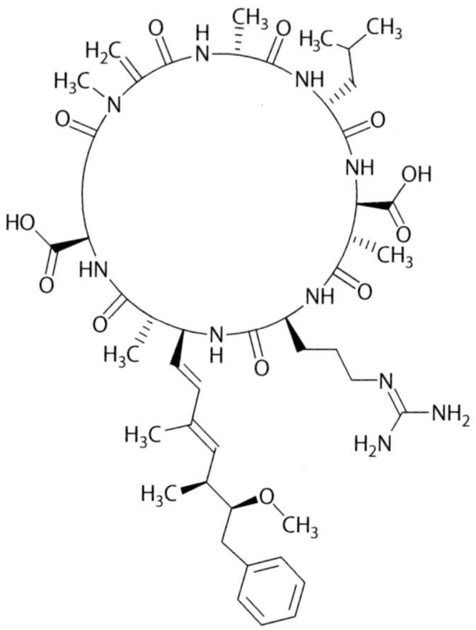

microcystin –LR

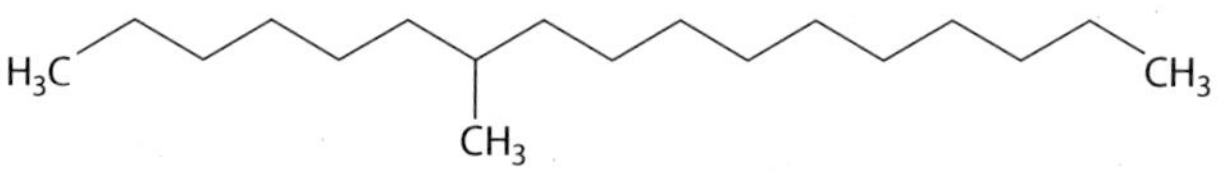

7-methyl heptadecane

SCYTONEMA JULIANUM MENEGHINI

Phylum: Cyanobacteria
Class: Cyanophyceae
Order: Nostocales
Family: Scytonemataceae
Distribution: Britain, Spain
Habitat: Freshwater habitats
Description: This free-floating, filamentous species forms dark mats and fixes atmospheric nitrogen.
Compounds and activities: The phosphoglycolipids, viz., phosphoglyco-analog of acetyl-acetylated sphingosine and glyco-analog of phosphatidyl glycerol, isolated from this species are potent inhibitors of platelet activating factor-induced platelet aggregation (Antonopoulon, 2005). Further, the extract of this species has also exhibited acetylene reduction activity.

SCYTONEMA BOHNERII SCHMIDLE

Phylum: Cyanobacteria
Class: Cyanophyceae
Order: Nostocales
Family: Scytonemataceae
Habitat: Freshwater ponds, streams, lakes
Description: This blue-green algal species has false-branching filaments.

Compounds and activities: The total lipid content (%) of this species was found to be 22.22 ± 0.32. The percent of total *n* saturated fatty acids (hexadecanoic acid) was 60%, and that of unsaturated fatty acids (9,12-octadecadienoic acid) was 40%. The polyunsaturated fatty acids (PUFAs) of this species play an important role in human metabolic pathways, particularly as specific precursors for prostaglandin E1, an important compound necessary for reducing inflammation and in treatment of heart disease (Sharathchandra and Rajashekhar, 2011). The bioactive compounds isolated from this species, viz., palmitic acid and linoleic acid, have been reported to possess antidiabetic properties.

CHROOMONAS PLACOIDEA BUTCHER, 1967

Phylum: Cryptophyta
Class: Cryptophyceae
Order: Pyrenomonadales
Family: Chroomonadaceae
Distribution: Mexico
Habitat: Lakes, marine habitats
Description: This cryptomonad (cryptophyte) species has large, elliptic, biflagellate free-swimming, motile cells that often form thick, mucilaginous, palmelloid, blue, blue-green, or green colonies. The cells possess a single chloroplast and pyrenoid with pigments, viz., phycocyanin and phycobiliprotein.
Compounds and activities: This species has yielded n-3 (α-linoleic acid, eicosapentaenoic acid, and docosahexaenoic acid) and n-6 (linolenic acid and arachidonic acid) (i.e., omega-3 and -6) polyunsaturated fatty acids (PUFAs) and 24-ethylcholesta-5,22E-dien-3β-ol. The omega-3 PUFAs improve both systolic and diastolic heart functions besides lowering inflammatory biomarkers. Further, PUFA 3 acids have been reported to reduce total and LDL cholesterol; thereby, they could be used as second-line therapy in the treatment of severe hypertriglyceridemia.

TRAPA BISPINOSA ROXB

Phylum: Magnoliophyta
Class: Magnoliopsida
Order: Myrtales
Family: Trapaceae
Common name: Water chestnut
Distribution: India
Habitat: Ponds, lakes, tanks
Description: This herbal species has a fluxuose stem, and the leaves (5 cm long) are rhomboid shaped and reddish purple. They are alternately arranged and crowded on the parts of the stem. Flowers are few, white, solitary, and axillary, and the fruit is white and obovoid.
Compounds and activities: The aqueous extracts of its fruits have also shown immunomodulating effects in animals, particularly in stimulating cellular and humoral response. The immunostimulatory property (stimulation of cellular and humoral response) of aqueous extracts of fruits of this species was confirmed by elevated neutrophil counts significantly ($p < 0.01$) compared to control values (Patel et al., 2010). The hydroalcoholic extracts of fruits of this species have shown analgesic activities in test animals, besides serving as a nervine tonic and improving learning and memory power. The crude extracts of this species have also shown antimicrobial, cytotoxic, and immunomodulatory properties.

WEDELIA CHINENSIS (OSBECK) MERR

Phylum: Magnoliophyta
Class: Magnoliopsida
Order: Asterales
Family: Asteraceae
Common name: Chinese wedelia
Distribution: China, India, Indonesia, Japan, Malaysia
Habitat: Paddy ridges, moist areas
Description: It is a tender, spreading, hairy herb with branches measuring about 50 cm. Leaves are oblong to oblong-lanceolate with a maximum length of 5 cm. Margins of the leaves are entire, and both surfaces of the

leaf are covered with sharp, pointed straight and stiff hairs. Heads (1 cm diameter) are yellow and stalked. Involucral bracts are oblong-ovate. Ray flowers, numbering about 10, are broad and spreading. Disk flowers, numbering about 20, are short, narrow, and pointed.

Compounds and activities: The antiproliferative compounds isolated from this species include indole-3-carboxyaldehyde, wedelolactone, luteolin, and apigenin possessing an androgen receptor (AR) suppression (inhibition) property. The 50% inhibition concentration values of these compounds (with respect to their ability to modulate the AR activation of transcription from prostate-specific antigen promoter in prostate cancer [PCa] cells) were found to be 34.9, 0.2, 2.4, and 9.8 μm, respectively (Lin et al., 2007).

HOTTONIA PALUSTRIS LINNAEUS

Phylum: Magnoliophyta
Class: Magnoliopsida
Order: Ericales
Family: Primulaceae
Common name: Featherfoil, water violet
Distribution: Europe, Northern Asia
Habitat: Calm shallow waters, ponds, ditches

Description: This perennial, submerged, and self-fertile plant grows to a height of 80 cm. While its basal roots are found buried in mud, its silvery shiny roots dangle freely in water. The leaves are pinnate and alternately arranged. Flowers are white-mauve and hermaphroditic.

Compounds and activities: As the flower extracts (essence) of this species contain vitamins A, C, and E and flavanoids, this healing herb has been used for treating several diseases, including heart problems.

GLINUS OPPOSITIFOLIUS (L) AUG. DC

Phylum: Magnoliophyta
Class: Magnoliopsida
Order: Caryophyllales
Family: Molluginaceae
Common name: Slender carpet weed
Distribution: Australia, Africa, India
Habitat: Damp sites

Description: It is a diffuse, slender, spreading or ascending, prostrate annual herb. Stems (40cm long) are slender, numerous, and dichotomously branched. Leaves (30 mm long) are opposite or alternate and are arranged in whorls of five, unequal, oblanceolate, or linear-lanceolate. Flowers are small white and are present in axillary fascicles of two.

Compounds and activities: Two pectin-type polysaccharides, GOA_1 and GOA_2 (acidic carbohydrate polymers), have been isolated from the aerial parts of this species. While GOA_1 possesses neutral sugars, arabinose (26.4 mol%), galactose (42.9 mol%), and arabinogalactans type I (AG I) and type II (AG II), GOA_2 contains rich galacturonic acid (68.3 mol%) along with rhamnose, arabinose, and galactose. Both GOA_1 and GOA_2 exhibited potent dose-dependent complement-fixating activities and induced chemoaxis of macrophage T cells and NK cells (Inngjerdingen et al., 2005). This species has also been used in traditional medicines and for the treatment of joint pain, inflammation, and wounds, besides possessing immunomodulatory, antipyretic, and antimalarial properties.

NYMPHAEA TETRAGONA GEORGI

Phylum: Magnoliophyta Class: Magnoliopsida
Order: Nymphaeles Family: Nymphaeceae
Common name: Pygmy water lily, small white water lily
Distribution: Asia, Europe, North America
Habitat: Freshwater habitats
Description: This dicot perennial herb has unbranched rhizomes, and the stem is erect and cylindrical stolons are absent. The leaves are petiolate and glabrous. The leaf blade (15 × 13 cm) is green to purple above and ovate to elliptic and green below. The floating flowers (4–8 cm diameter) have 10–15 white petals.
Compounds and activities: The bioactive compound, geranin (a hydrolizable tannin), isolated from this species has been reported to reduce blood pressure, decrease serum lipids, produce liver necrosis, and modulate immunoresponses. Further, it possesses anticancer, antimutagenic, and enhanced blood-clotting properties. The stem of this species has been shown *in vitro* to be not only a potent antioxidant, but also able to inhibit the expression of collagen-depleting collagenase (matrix metalloproteinase-1 (MMP-1)) when applied to human fibroblast cells. Further, the extract of this species has been reported to inhibit the growth of unwanted algal blooms in freshwater systems (Li and Hou, 2007).

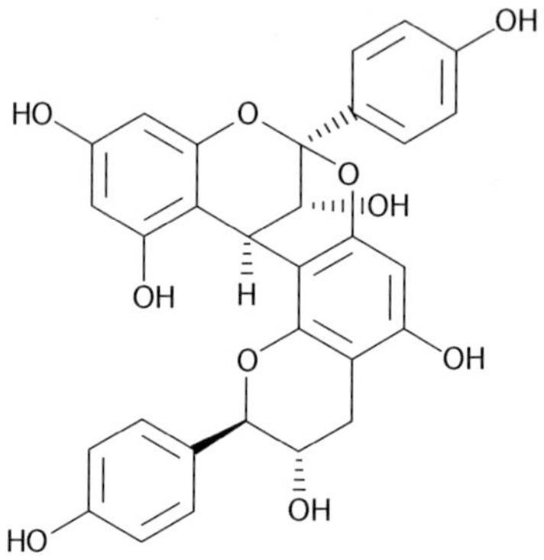

geranin

NYMPHAEA ODORATA AITON

Phylum: Magnoliophyta Class: Magnoliopsida
Order: Nymphaeles Family: Nymphaeceae
Common name: Fragrant water lily, white water lily
Distribution: North America, Central America, Canada
Habitat: Shallow lakes, ponds, slow-moving waters
Description: This species is rooted from a branched rhizome from which the long petioles and smooth floating leaves arise. The floating flowers are radially symmetrical with white petals and yellow stamens.
Compounds and activities: Several phenolic compounds have been isolated from this species (Zhang et al., 2003). The bioactive compounds of this species include nymphaeoside A, icariside E4, kaempferol 3-O-α-L-rhamnopyranoside(afzelin), quercetin 3-O-α-L-rhamnopyranoside, myricetin 3-O-α-L-rhamnopyranoside(myricitrin), quercetin-3-O-(6″-O-acetyl)

β-D-galactopyranoside, myricetin-3-O-β-D-galactopyranoside, and myricetin-3-O-(6″-O-acetyl) β-D-galactopyranoside. The flavonoid glycosides and lignans of this species possess inhibitory activity against fatty acid synthase, a multienzyme protein for catalyzing fatty acid synthesis. This process would help prevent obesity in humans (Zhang et al., 2003).

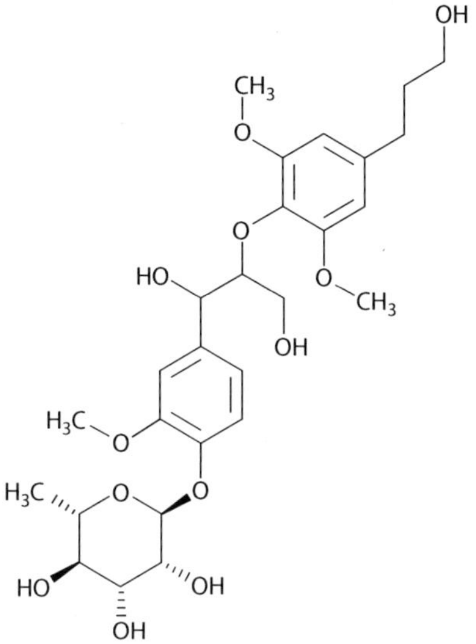

nymphaeoside A

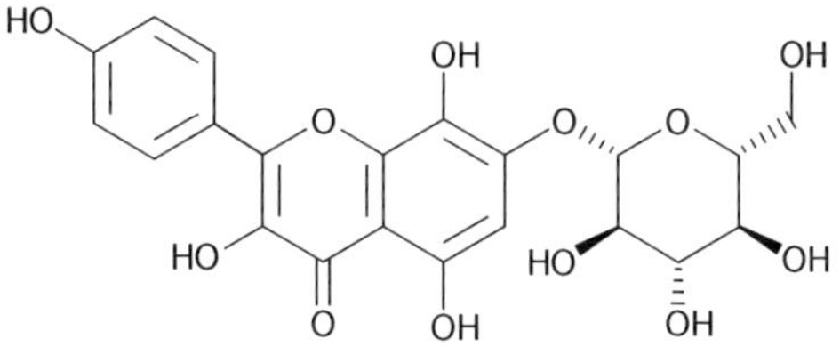

icariside E4

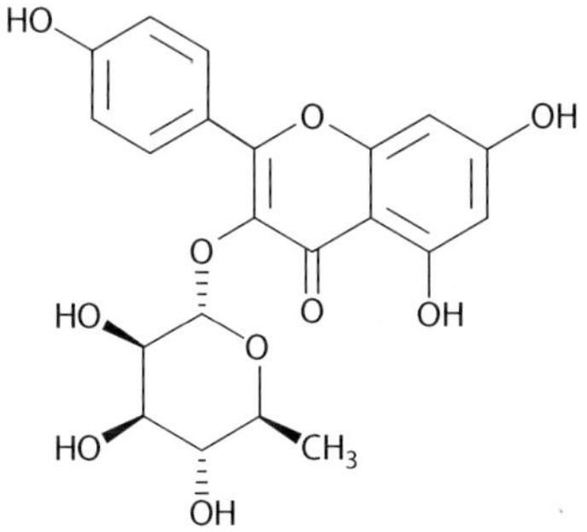

kaempferol 3-O-α-L-rhamnopyranoside(afzelin)

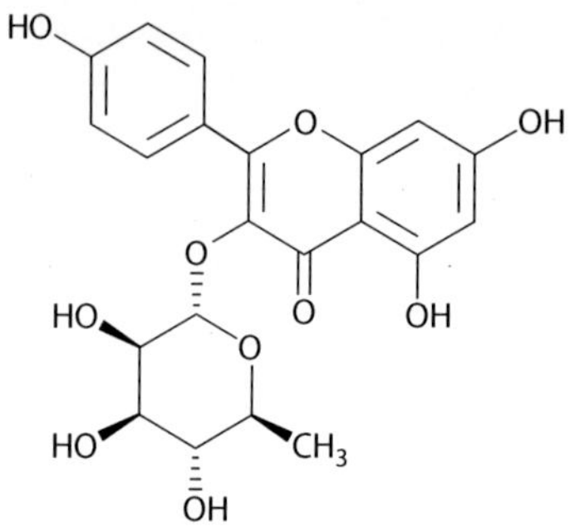

quercetin 3-O-α-L-rhamnopyranoside

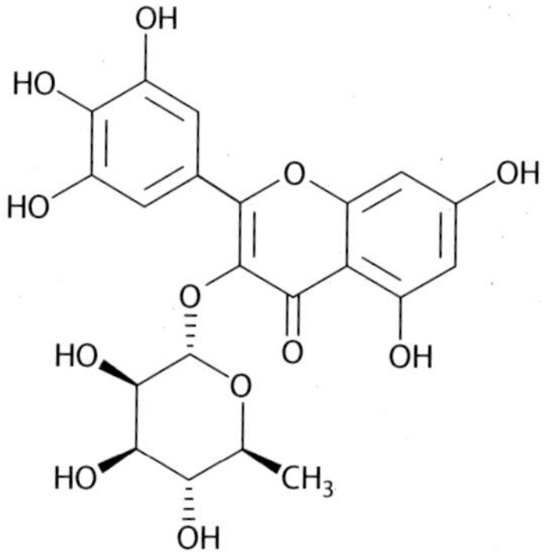

myricetin 3-O-α-L-rhamnopyranoside(myricitrin)

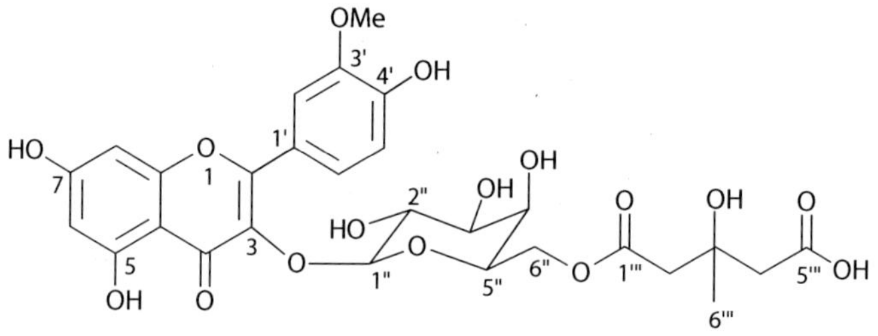

quercetin-3-O-(6"-O-acetyl) β-D-galactopyranoside

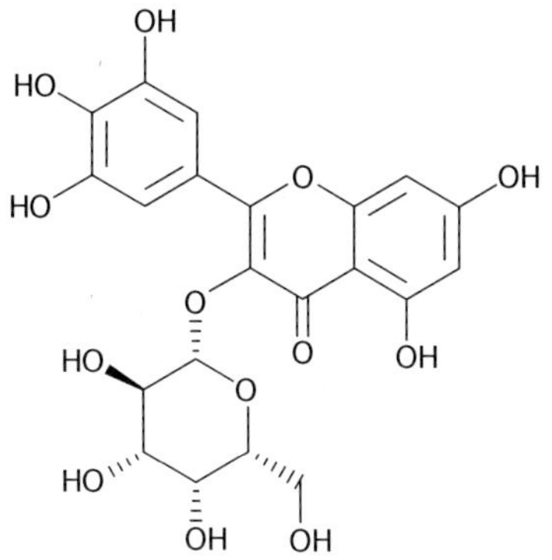

myricetin-3-O-β-D-galactopyranoside

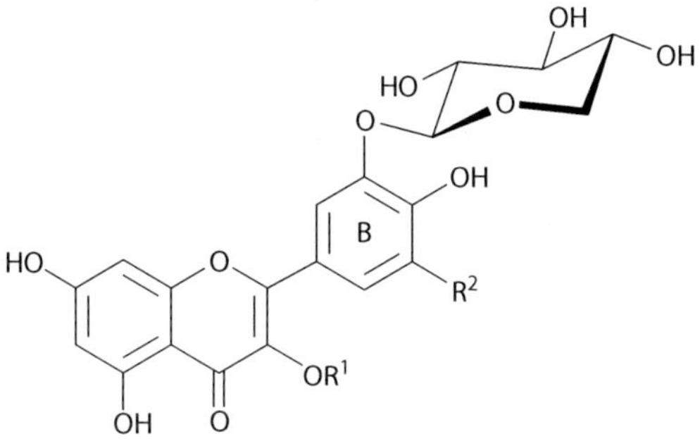

myricetin-3-O-(6"-O-acetyl) β-D-galactopyranoside

ECLIPTA ALBA LINNAEUS

Phylum: Magnoliophyta
Class: Magnoliopsida
Order: Asterales
Family: Asteraceae
Common name: False daisy

Distribution: India, China, Thailand, Brazil
Habitat: Paddy fields, moist places
Description: The plant is an erect, slender, and roughly pubescent herb growing to a maximum height of 60 cm. Leaves are opposite, sessile, narrowly lanceolate, toothed, or nearly entire. Floral heads, which are 8 mm in diameter, are solitary, and florets are white. Roots are well developed, grayish, and cylindrical.
Compounds and activities: This species contains three major coumestans, viz., norwedelolactone, wedelolactone, and demethylwedelolactone. Other compounds present in this species are flavonoids, steroids, triterpenoids, polyacetylenes, and thiophene derivatives. The leaf extract of this species has shown significant hepatoprotective activity against paracetamol-induced rat (Prabu et al., 2011). Experimented rats (with a single dose of paracetamol (2 g/kg) and leaf extract (500 mg/kg) for 7 days) showed that the values of serum protein, serum bilirubin, serum cholesterol, and serum triglycerides were less than those of animals treated with paracetamol and silymarin (standard hepatoprotective agent). This species has also been used in traditional medicines for the treatment of edema, joint pain, fever, and enlarged liver, besides having uretic and memory-enhancing properties (Prabu et al., 2011).

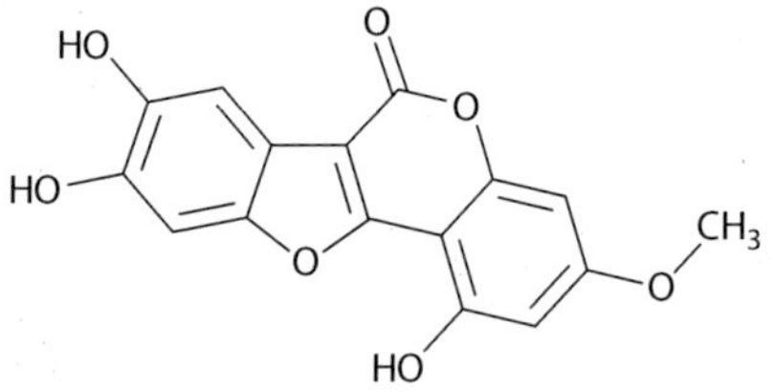

wedelolactone

AMMANIA BACCIFERA LINNAEUS

Phylum: Magnoliophyta
Class: Magnoliopsida
Order: Myrtales
Family: Lythraceae

Common name: Acrid weed, tooth cup
Distribution: United States
Habitat: Swampy and damp sites
Description: It is an erect, branched, smooth, slender, annual purplish herb growing to a maximum height of 50 cm with four-angled stems. Leaves are narrow-oblong, oblance shaped with 4 cm length each. Flowers are small, 1 mm long, greenish or purple, and borne in dense clusters in the leaf axis.
Compounds and activities: The ethanol extract of this species possesses an antisteroidogenic property. At doses of 100, 200, and 400 mg/kg body weight, this extract was found to arrest the normal estrus cycle at dioestrus phase and significantly reduced the weight of ovaries. The extract also significantly inhibited the activity of Δ5-3β-hydroxyl steroid dehydrogenase (Δ5-3β-HSD) and glucose-6-phosphate dehydrogenase (G-6-PD) enzymes involved in ovarian steroidogenesis (Dhanapal et al., 2005). Leaves or ashes of plant mixed with oil are applied to cure herpetic eruptions, and fresh leaves are used for treating skin diseases.

MENTHA ARVENSIS LINNAEUS

Phylum: Magnoliophyta
Class: Magnoliopsida
Order: Lamiales
Family: Lamiaceae
Common name: Field mint, wild mint, corn mint
Distribution: Europe, Asia, Siberia, North America
Habitat: Wet grasslands near rivers and lakes
Description: It is a perennial herbaceous plant growing to a maximum height of 1 m. Leaves (7 × 2 cm) are simple and are in opposite pairs. They are hairy with coarsely serrated margins. Flowers (4 mm long) are purple, pink, or white and are present as clusters on the stem.
Compounds and activities: The extracts of this species showed highly significant activity against gastric ulcers. A decrease in gastric secretion and ulcer index was recorded among the treated groups, i.e., petroleum ether (53.4%), chloroform (59.2%), aqueous (67.0%), and in standard drug (68.7%), when compared to the negative control (Londonkar and Poddar, 2009).

JUNCUS SUBULATUS FORSSKAL

Phylum: Magnoliophyta
Class: Liliopsida
Order: Poales
Family: Juncaceae
Common name: Common wood rush, somerset rush
Distribution: Britain, Scotland, Mediterranean-Atlantic
Habitat: Freshwater pools, brackish water
Description: This medium tall, rhizomatous plant forms large dominant patches. Leaves, which are alternate and rosette, are entire with smooth margins. Flowers are hermaphroditic and green or brown.
Compounds and activities: The volatile oil of the tubers of this species contains 17 compounds, including 3 major compounds, viz., bergamal, 2-methyl-4-heptanone, and lauric acid. The ethyl acetate of this species showed five 9,10-dihydrophenanthrenes, luteolin, β-sitosterol, and stigmasterol. The n-butanol extract of this species has two flavonoid glycosides and quercetin. All the above extracts showed hepatoprotective activity in the following order: volatile oil extract > ethyl acetate extract > n-butanol extract (Razik et al., 2009).

CYPERUS ROTUNDUS LINNAEUS

Phylum: Magnoliophyta
Class: Liliopsida
Order: Poales
Family: Cyperaceae
Common name: Coco-grass, nut grass
Distribution: Africa, Europe, France, Austria, Asia
Habitat: Banks of rivers and streams, crop fields
Description: This perennial species grows to a maximum height of 150 cm. Leaves (20 cm long) sprout in ranks of three from the base of the plant. Hermaphroditic flowers have three stamina and three stigma, carpel. Rhizomes are blackish, hard, and fragrant tubers.
Compounds and activities: The alcoholic extract (280 mg/kg) of the rhizome of this species showed significant reduction in serum lipid profile. At this extract concentration, the values of serum total cholesterol, serum low-density lipoprotein, serum triglyceride, and serum high-density lipoprotein were found to be 239.44 ± 4.028, 119.37 ± 2.60, 276.53 ± 9.70, and 64.76 ± 0.62 mg/dl (Chandratre et al., 2011).

TYPHA ANGUSTATA BORY & CHAUB

Phylum: Magnoliophyta
Class: Liliopsida
Order: Poales
Family: Typhaceae
Common name: Lesser Indian reed mace
Distribution: Pantropical, Pakistan
Habitat: Shallow ponds, riverbanks
Description: It is a perennial plant growing to a maximum height of 2 m. The stem is erect, and the leaves are thick, with a maximum width of 12 mm. Male and female flowers develop in complex spikes on the same vertical stem.
Compounds and activities: The bioactive compounds isolated from this species include bacosides, bacopasides, and bacopasaponins. Its (2S)-naringenin inhibits proliferation of vascular smooth muscle cells (VSMCs), thereby avoiding myocardial infarction (Lee et al., 2012).

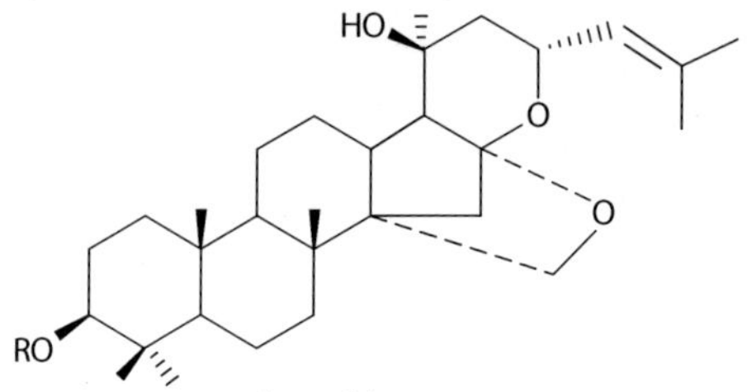

bacosides

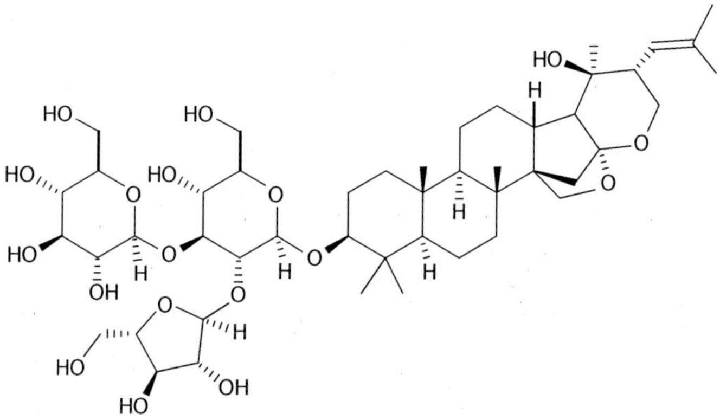

bacopasides

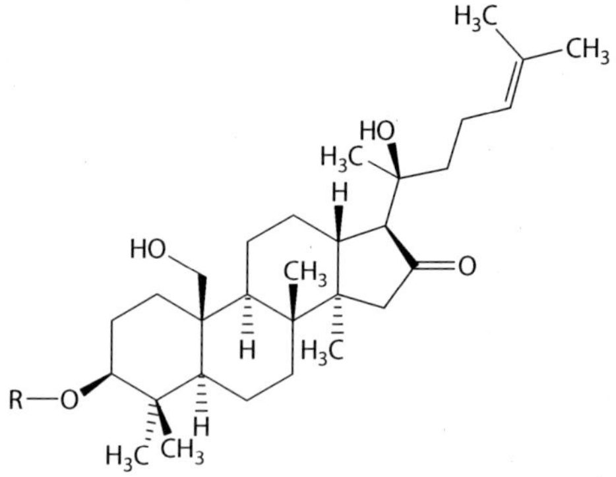

bacopasaponins

VALLISNERIA SPIRALIS LINNAEUS

Phylum: Magnoliophyta
Class: Liliopsida
Order: Alismatales
Family: Hydrocharitaceae
Common name: Eelgrass, tapegrass
Distribution: Southeast Asia, South Europe
Habitat: Lakes, ponds, marshes, streams
Description: It is an evergreen perennial species. Leaves are narrow and linear and are pale green to reddish. Leaves measure 100 m long and 2 cm wide. Flowers are dioecious and are carried on long spiral stalks.
Compounds and activities: The extract of this species has yielded bioactive compounds, viz., 2-ethyl-3-methylmalemide, dihydroactinidiolide, 4-oxo-β-ionone, 3-hydroxy-5,6-epoxy-β-lonone, loliolide, and 6-hydroxy-3-oxo-α-lonone. All these compounds showed antialgal properties and inhibitory effects on *Microcystis aeruginosa* (Xian et al., 2006).

SAGITTARIA TRIFOLIA LINNAEUS

Phylum: Magnoliophyta Class: Liliopsida
Order: Alismatales Family: Alismataceae
Common name: Three-leaf arrowhead
Distribution: Iran, Pakistan, China, Malaysia, Japan, Australia
Habitat: Paddy fields, ponds, lakes
Description: This glabrous herb has thick stolons bearing tubers at the tips. Leaves (15 × 6 cm) are emerged, upright, and sagittate. The blade is ovate or linear-lanceolate. Inflorescence is an unbranched scapose raceme, 50 cm long. Monoecious, white flowers (2 cm across) are in six whorls of five flowers each.
Compounds and activities: From the tuber extracts of this species four bioactive diterpene ketones, viz., trifoliones A–D; two diterpene glucosides, viz., sagittariosides *a* and *b*; nitroethylphenol glycoside, arabinothalictoside; and six known diterpenes have been isolated. Among them, the compounds trifoliones A–D showed inhibitory effects on the histamine release from the mast cells induced by compound 48/80 and calcium ionophore A-23187 (Yoshikawa et al., 1996).

References

Abdo, S.M., M.H. Hetta, F.A. Samham, R.A.S. Eldin, and G.H. Ali. 2012. Phytochemical and antibacterial study of five freshwater algal species. *Asian J. Pl. Sci.*, 11:109–116.

Abul-Hajj, Y. Staba, E.J. Su, and K. Lee. 1972. *Aquatic plants from Minnesota. Part 3. Antimicrobial effects.* Water Resources Research Centre (WRRC) Bulletin 48.

Afifi, M.S., A.L. Kansoh, and M.M. Abdulla. 2009. Flavanoids and bioactivity of *Myriophyllum spicatum* L. (Haloragaceae) growing in Egypt (Abstract, online). http://www.pharma.cu.edu.eg/Files/BFOPCU/Volums/TOC_2009_V47_Issue_1/12_01_2009.html

Akinjogunla, J.O., C.S. Yah, N.O. Egliafona, and F.O. Ogbemudia. 2010. Antibacterial activity of *Nymphaea lotus* (Nymphaceae) leave extract against MRSA and VRSA from chemical samples. *Elect. J. Environ. Agric. Food Chem.*, 58:1064–1073.

Al-Wathnani, H., I. Ara, R.R. Tahmaz, T.H. Al-Dayel, and M.A. Bakir. 2012. Bioactivity of natural compounds isolated from cyanobacteria and green algae against human pathogenic bacteria and yeast. *J. Med. Plant Res.*, 6(18):3425–3433.

Amaro, H.M., A.C. Guedes, and F.X. Malcata. 2011. Antimicrobial activities of microalgae—An invited review. In *Science against microbial pathogens: Communicating current research and technological advances*, ed. A. Mendez-Vilas, 1272–1280. Formatex Research Center, Badajoz, Spain.

Anthoni, U., P.H. Nelsen, L.S. Hansen, S.W. Anderson, and C. Christophersen. 1987. Charamin, a quarternary ammonium ion antibiotic from the green algae *Channa globularis. J. Org. Chem.*, 52:694–695.

Antonopoulon, S., T. Nomikos, A. Oikonomou, A. Kyriacou, M. Andriolis, E. Fragopoulou, and A. Pantazidou. 2005. Characterisation of bioactive glycolipids from *Sytonema julianum* (Cyanobacteria). *Comp. Biochem. Physiol. B Biochem. Mol. Biol.*, 140:219–231.

Araki, H., M. Inou, and T. Katoh. 2003. Total synthesis and absolute configuration of octillions A and B, novel and potent antitumor agents from a freshwater plant. *Org. Lett.*, 5:3903–3906.

Arallado, E.C., and M.O. Osi. 2010. Hematinic activity of *Alternanthera sessilis* (Amaranhtaceae) in mice and rats. *E-Int. Sci. Res. J.* 2:110–117.

Armando, C.C., and O.O. Dennis. 2010. Preliminary phytochemical and antimicrobial evaluation of the fresh and dried whole plant extracts from *Commelina benghalensis. Rev. Colombiana Cien. Anim.*, 2:104–116.

Asha Devi, S., and D. Ganjewala. 2009. Antimicrobial activity of *Acorus calamus* (L) rhizome and leaf extract. *Acta Biol. Szegediensis*, 53:45–49.

Atawodi, S.E.O., and G.S. Onaolapo. 2010. Comparative in vitro antioxidant potential of different parts of *Ipomoca asarifolia*, Roemer & Schultes, *Guiera senegalensis*, J.F. Gmel and *Anisopus mannii* N.E. Brown. *Braz. J. Pharm. Sci.*, 46:245–248.

Azamai, E.S.M., S. Sulaiman, S.H.M. Habib, W.Z.W. Ngak, and Y.A.M. Yusof. 2009. *Chlorella vulgaris* triggers apoptosis in hepatocarcinogenesis-induced rats. *J. Zhejiang Univ. Sci. B* 10:14–21.

Bauer, N. 2011. Dynamic of allelopathically active polyphenolic substances of *Myriophyllum verticillatum* L. and factors influencing allelopathic effects on phytoplankton. Dr. Rer. Nat. dissertation, Hambolt University, Berlin.

Behery, F.A., Z.E. Naeem, G.T. Maatooq, M.M. Amer, and A.F. Ahmed. 2012. A novel antioxidant phenanthrenoid dimer from *Junctus acutus* L. *Nat. Prod. Res.*, 27:155–163.

Bero, J., M. Frederich, and J.Q. Leclercq. 2009. Antimalarial compounds isolated from plants used in traditional medicine. *J. Pharm. Pharmacol.*, 61:1401–1433.

Blom, J.F., and J. Pernthaler. 2010. Antibiotic effects of three strains of chrysophytes (*Ochromonas* and *Poterioochromonas*) on freshwater bacterial isolates. *FEMS Microbiol. Ecol.*, 71:281–290.

Bokesh, H.R., B.R. Okeefe, T.C. McKee, L.K. Pennell, G.M.L. Patterson, R.S. Gardella, R.C. Sowder, J. Turpin, K. Watson, R.W. Buckhert, and M.R. Boyd. 2003. A potent novel anti-HIV protein from the cultured cyanobacterium, *Scytonema varium. Biochemistry*, 42:2578–2584.

Boopathy, N.S., and K. Kathiresan. 2010. Anticancer drugs from marine flora: An overview (online). *J. Oncol.*, 2010:214186.

Borchardt, J.R., D.L. Wyse, C.C. Sheaffer, K.L. Karuppi, R.G. Fulcher, N.J. Ehlke, D.D. Biesboer, and R.F. Bey. 2009. Antioxidant and antimicrobial activity of seed from plants of Mississippi river basin. *J. Med. Pl. Res.* 3:707–718.

Bream, A.S., M.I. Hassan, M.A. Fouda, and T.M. El-Sheikh. 2009. Toxicity and repellent activity of *Phragmites australis* extracts against the mosquito vector *Culex pipiens. Tunisian J. Plant. Prot.*, 4:157–172.

Buznego, M.T., and H. Perez-Saad. 2006. Behavioral and antiepileptic effect of acute administration of the extract of the aquatic plant *Echinodorus berteroil* Fassett (upright burhead). *Epilepsy Behav.*, 9:40–45.

Cao, Z.G., J.H. Liu, S.W. Zhou, W. Wu, C.P. Yin, and J.Z. Wu. 2004. The effects of active constituents of *Alisma orientalis* on renal stone formation and bikunin expression in rat urolithasis model. *Zhonghua Yi Xue Za Zhi*, 84:1276–1279.

Challam, M., B. Roy, and V. Tandon. 2012. In vitro antihelmintic efficacy of *Carex baccans* (Cyperaceae): Ultrastructural, histochemical and biochemical alterations in the cestode, *Raillietina echinobothrida. J. Par. Dis.*, 36:81–86.

Chandratre, R.S., S. Chandarana, and M.A. Mengi. 2011. Lipid lowering activity of alcoholic extract of *Cyperus rotundus. Int. J. Res. Pharm. Chem.*, 1:1042–1045.

Chang, C.-I., and Y.-H. Kuo. 2007. Oleanone type triterpenes from *Ludwigia octovalis. J. Asian Nat. Proc. Res.*, 9:67–72.

Chaudhary, V., R. Prasanna, and A.K. Bhatnagar. 2012. Modulation of fungicidal potential of *Anabaena* strains by light and temperature. *Folia Microbiol.* 57:199–208.

Choosawad, D., U. Leggat, A. Phongdara, W. Chotigeat, and C. Dechsukhum. 2005. Anticoagulant and antitumor activity of crude fucoidan extracted from aquatic plant, *Utrichlaria aurea.* Lour. *Songklanakarin. J. Sci. Technol.*, 27(Suppl. 3):799–807.

Chowdhury, N.S., M.B. Alam, A.S.M.T. Haque, M.R. Zohan, H.E. Mazumder, and M.E. Haque. 2011. In vitro free radical scavenging and thrombolytic activities of Bangladesh aquatic plant, *Aponogeton undulatus* Roxb. *Global J. Pharmacol.*, 5:27–32.

Chung, Y.M., Y.H. Lan, T.L. Hwang, and Y.L. Leu. 2011. Anti-inflammatory and antioxidant components from *Hygroryza aristata. Molecules*, 16:1917–1927.

Corona, A.H., I. Nieves, M. Meckes, G. Chamorro, and B.L. Barron. 2002. Antiviral activity of *Spirulina maxima* against herpes simplex virus type 2. *Antiviral Res.*, 56:279–285.

Cui, H.H.H., Z.Zy. Wang, Z.Z.Q. Wang, R.R.M. Li, and Y.Y.H. Kang. 2005. Study on the optimal harvesting time of *Rumex gmelini* by analyzing the contents of principal components. *China J. Chin. Mater. Med.*, 30:110–115.

Das, B., J. Kundu, S.C. Bachar, M.A. Uddin, and J.K. Kundu. 2007. Antitumor and antibacterial activity of ethyl acetate extract of *Ludwigia hyssopifolia* Linn. and its active principle piperine. *Pak. J. Pharm. Sci.*, 20:128–131.

Dhanapal, R., S. Kavimani, V.S. Bhyrapurmatha, M. Gupta, and S.K. Basu. 2005. Antisteroidogenic activity of ethanol extract of *Ammania baccifera* (L) whole plant in female albino mice ovaries. *Iranian J. Pharmacol. Ther.*, 4:43–45.

Ding, T., T. Jiang, J. Zhou, L. Xu, and Z.M. Gao. 2010. Evaluation of antimicrobial activity of endophytic fungi from *Camptotheca acuminata* (Nyssaceae). *Genet. Mol. Res.*, 9:104–112.

Drapean, C. 2003. Anti-inflammatory and antidepressant properties on the blue green algae *Aphanizomenon flo-saquac*. *Comp. Health Prac. Rev.*, 8:161–162.

Elzaawely, A.A., and S. Tawata. 2012. Antioxidant capacity and phenolic content of *Rumex dentatus* L. grown in Egypt. *J. Crop Sci. Biotechnol.*, 15:59–64.

Elzaawely, A.A., T.D. Xuan, and S. Tawata. 2005. Antioxidant and antibacterial activities of *Rumex japonicus*. Houtt, aerial parts. *Biol. Pharm. Bull.*, 28:2225–2230.

Fang, J.B., W. Jia, W.Y. Gao, Z. Yao, J. Teng, A.H. Zhao, and H.Q. Duan. 2007. Antitumor constituents from *Alternanthera philoxeroides*. *J. Asian Nat. Prod. Res.*, 9:511–515.

Fossen, T., and O.M. Anderson. 1999. Anthocyanins from blue flowers of the African water lily, *Nymphaea coerula*. *Phytochemistry*, 50:1185–1188.

Ghasemi, Y., A. Moradian, A. Mohagheghazadeh, S. Shokravi, and M.H. Morowvat. 2007. Antifungal and antibacterial activity of the microalgae collected from paddy fields of Iran: Characterization of antimicrobial activity of *Chroococcus dispersus*. *J. Biol. Sci.*, 7:904–910.

Ghazala, B., B. Naila, and M. Shameel. 2009. Phycochemistry and bioactivity of ten freshwater algae from Pakistan. *Int. J. Algae* 11:84–98.

Ghazala, B., B. Naila, M. Shameel, S. Shahzad, and S.M. Leghari. 2004a. Phytochemistry and bioactivity of two stonewort algae (Chlorophyta) of Sindh. *Pak. J. Bot.*, 36:733–743.

Ghazala, B., M. Shameel, M.I. Choudhary, S. Shahzad, and S.M. Leghari. 2004b. Phytochemistry and bioactivity of *Tetraspora* (Volvocophyta) from Sindh. *Pak. J. Bot.*, 36:531–547.

Ghobrial, M.G., S. Omar, and T. Adnan. 2007. Potential for finding new bioactive agents from selected aquatic plant extracts against enteric bacteria. *Egypt. J. Nat. Toxins*, 42:12–25.

Gibbs, A., C. Green, and V.M. Doctor. 1983. Isolation and anticoagulant properties of polysaccharides of *Typha angustata* and *Daemonorops* species. *Thromb. Res.*, 15:97–108.

Gosmann, G., S. Gattuso, M. Gattuso, R. Fenner, E.F. Pacheeo, A. Ferraz, L.A. Savi, C.R.M. Barard, C.M.O. Simoes, M. Sortino, S. Zacchino, C. Gnerre, B. Testa, and S.M.K. Rates. 2003. Botanical (morphological), micrographic, chemical and pharmacological characteristics of *Pfaffia* species (Amaranthaceae) native to South Brazil. *Braz. J. Pharm. Sci.*, 39:141–147.

Govindan, G., T.G. Sambandan, M. Govindan, A. Sinskey, J. Vanessendelft, I. Adenan, and C.K. Rha. 2007. A bioactive polyacetylene compound isolated from *Centella asiatica Planta Med.*, DOI 10.1055/s-2007.981521.

Greca, D., L. Previtera, and A. Zaralli. 2009. Structures of new phenyl phoralene-related compounds from *Eichnornia crassipes*. *Tetrahedron Lett.*, 30:1–3.

Greca, M.D., A. Fiorentino, M. Isidori, P. Monaco, F. Temussi, and A. Zarelli. 2001. Antialgal furano-diterpenes from *Potamogeton natans*. *Phytochemistry*, 58:299–304.

Gutierrez, R.M.P., and G.R.V. Solis. 2007. Smooth muscle relaxing properties of the hexanic extract of freshwater algae *Oscillatoria limnetica* and *Hydrodiclyon reticulatum*. *Boletin Latinoam. Caribe Plantas Med. Aromatics*, 6:30–35.

Hamkrasri, A., S. Chankhamhaengdecha, and P. Damrongphol. 2011. Bioactivities of three freshwater submersed macrophytes from Bueng Boraphet. In *Proceedings of the 37th Congress on Science and Technology*, Thailand, pp. 1–4.

Hasan, M.F., R. Das, A. Khan, M.S. Hossain, and M. Rahman. 2009. The determination of antibacterial and antifungal activities of *Polygonam hydropiper* (L) root extract. *Adv. Biol. Res.*, 3:53–56.

Hromadkova, Z., J. Hirsch, and A. Ebringerova. 2010. Chemical evaluation of *Fallopia* species leaves and antioxidant properties of their non-cellulose polysaccharide. *Chem. Mater. Sci.*, 64:663–672.

Huang, D.J., H.J. Chen, C.D. Lin, and Y.H. Lin. 2005. Antioxidant and antiproliferative activities of water spinach (*Ipomoca aquatica* Forsk.) constituents. *Bot. Bull. Acad. Sin.*, 46:99–106.

Inngjerdingen, K.T., S.C. Debes, M. Inngjerdingen, S. Hokputsa, S.E. Harding, B. Rolstad, T.E. Michaelsen, D. Diallo, and B.S. Paulsen. 2005. Bioactive pectic polysaccharides from *Glinus oppositifolius* (L.) Aug. DC, a Malian medicinal plant, isolation and partial characterization. *J. Ethnopharmacol.*, 101:204–214.

Jalalpure, S.S., N. Agrawal, M.B. Patil, R. Chimkode, and A. Tripathi. 2008. Antimicrobial and wound healing activities of leaves of *Alternanthera sessilis* Linn. *Int. J. Green Pharm.*, 2:141–144.

Jani, M., S. Shan, and S. Prajapati. 2012. Antibacterial screening and qualitative phytochemical estimation of selected aquatic plants. *Adv. Biol. Res.*, 6:19–23.

Jawaid, T., R. Gupta, and Z.A. Siddiqui. 2011. A review on herbal plants showing antidepressant activity. *Int. J. Pharm. Sci. Res.*, 2:3051–3060.

Jeyashree, S., J. Rathinamala, and P. Lakshmanaperumalsamy. 2011. Antimicobial activity of *Vetiveria zizaniodes* against some pathogenic bacteria and fungi. *Med. Plants Int. J. Phytomed. Relat. Indust.*, 3:151–156.

Jiang, Z., X. Zhang, F. Zhang, N. Liu, F. Zhau, J. Zhou, and J. Chen. 2006. A new triterpene and anti-hepatitis B virus active compounds from *Alisma orientalis. Planta Med.*, 72:951–954.

Joseph, A.R., A.V. Ajisha, S. Kumar, and S. Sujatha, 2011. Effect of bioactive compounds and their pharmaceutical activities of *Sida cordifoliai* (Linn). *Int. J. Biomed. Res.*, 2:1038–1042.

Jovin, E., N. Simin, D. Orcic, K. Balog, I. Beara, M. Lesjak, and N.M. Dukic. 2011. Antioxidant and anti-inflammatory properties of *Rumex patientia* L. *Planta Med.*, 77–157.

Kang, H., A. Krunic, and J. Orjala. 2012. Sanctolide A, a 14-membered PK-NRP hybrid macrolide from the cultured cyanobacterium *Oscillatoria sancta. Tetrahedron Lett.*, 53:3563–3567.

Kang, Y.H., Z.Y. Wang, J.K. Li, and L. Liu. 1996. Isolation and identification of two anthraquinones from *Rumex gmelini* Turz. *China J. Chin. Mater. Med.*, 21:741–742.

Kartal, M., I. Orhan, M. Abu-Asaker, F.S. Senol, T. Atici, and B. Sener. 2009. Antioxidant and anticholinesterase assets and liquid–chromatography–mass spectrometry preface of various freshwater and marine microalgae. *Pharm. Mag.*, 5:291–297.

Kaushik, D., A. Tripathi, R. Tripathi, M. Ganachari, and S.A. Khan. 2009. Anticonvulsant activity of *Bacopa monniera* in rodents. *Braz. J. Pharm. Sci.*, 45(4).

Kawahara, T., and T. Okino. 2012. Chlorosulfolipids. In *Studies in Natural Chemistry*, ed. A. Rahman, Vol. 36, Ch. 7, 219–248. Elsevier, UK.

Khalid, M.N., M. Shameel, B. Ghazala, and V.U. Ahmad. 2010. The bioactivity and phytochemistry of two stonewort algae (charophycota) from Sindh. *Proc. Pakistan Acad. Sci.*, 47:205–214.

Khaliq-uz-zaman, S.M., S. Shameel, M. Shameel, S.M. Leghari, and V.U. Ahmad. 1998. Bioactive compounds in *Chara corallina* var. *Wallichii* (A. Br.) R.D. Wood (Charophyta). *Pak. J. Biol.*, 30:19–31.

Khanavi, M., M. Moshteh, A. Manayi, M.R.S. Ardekani, M. Vazirian, Y. Ajani, and S.N. Ostad. 2011. Cytotoxic activity of *Lythrum salicaria. Res. J. Biol. Sci.* 6:55–57.

Kim, H. 2005. Characterisation of bioactive compounds in essential oils, fermented anchovy sauce and edible plants and induction of phytochemicals from edible plant using methyl jaspmonate (MeJA) and chitosan. PhD thesis, Clemson University.

Kim, K.H., E. Moon, H.K. Kim, J.Y. Oh, S.Y. Kim, S.U. Choi, and K.R. Lee. 2012. Phenolic constituents from the rhizomes of *Acorus gramineus* and their biological evaluation on antitumor and anti-inflammatory activities. *Bioorg. Med. Chem. Lett.*, 22:6155–6159.

Kimura, Y., and H. Okuda. 2001. Resveratrol isolated from *Polygonum cuspidatum* root prevents tumor growth and metastasis to lung and tumor-induced neovascularization in Lewis lung carcinoma–bearing mice. *J. Nutr.*, 131:1844–1849.

Kittakoop, P., S. Wanasith, and P. Watts. 2001. Potent antiviral potamogetonyde and potamogetenol, new furanoid labdane diterpenes from *Potamogeton malaianus*. *J. Nat. Prod.*, 64:385–388.

Kobayashi, J., M. Sekiguchi, S. Shimamoto, H. Shigemori, and A. Ohsaki. 2000a. Echinophyllins C–F, new nitrogen-containing clerodane diterpenoids from *Echinodorus macrophyllus*. *J. Nat. Prod.*, 63:1576–1579.

Kobayashi, J., M. Sekiguchi, S. Shimamoto, H. Shigemori, and A. Ohsaki. 2000b. Echinophyllins A and B, novel nitrogen-containing clerodane diterpenoids from *Echinodorus macrophyllus*. *Tetrahedron Lett.*, 41:2939–2943.

Kok, Y.Y., W.L. Chu, S.M. Phang, S.M. Mohamed, R. Naidu, P.J. Lai, S.N. Ling, J.W. Mak, P.K. Lim, P. Balraj, and A.S. Khoo. 2011. Inhibitory activities of microalgal extracts against Epstein-Barr virus DNA release from lymphoblastoid cells. *J. Zhejiang Univ. Sci. B*, 12:335–345.

Kokpol, U., T. Higa, and P. Sukpanyalert. 2000. Bioactive compounds of *Sphaeranthus africanus* Linn. Part II. In *26th Congress of Science and Technology*, Thailand.

Kshirsagar, A.D., K.G. Ingale, N.S. Vyawahare, and V.S. Thorve. 2010. *Hygrophila spinosa*: A comprehensive review. *Plant Rev.*, 4:167–171.

Kumar, M., M.K. Tripathi, A. Srivastava, G. Nath, and R.K. Asthana. 2012. A comparative study of antibacterial activity of brackish water and freshwater cyanobacterial strains. *Asian J. Exp. Biol. Sci.*, 3:548–552.

Lee, J.J., H. Yi, I. Kim, Y. Kim, N.X. Nhiem, Y.H. Kim, and C.S. Myung. 2012. (2S)-Naringenin from *Typha angustata* inhibits vascular smooth muscle proliferation via a Go/G12 arrest. *J. Ethnopharmacol.*, 139:873–878.

Lee, S.H., A.D. Kim, M.C. Kang, J.B. Lee, and Y.J. Jeon. 2009. Potential antioxidant activities of enzymatic digests from freshwater microalgae *Pediastrum duplex* and *Dactylococcopsis fascicularis*. *Algae*, 24:169–177.

Leu, Y.L., T.L. Hwang, J.W. Hu, and J.Y. Fang. 2008. Anthroquinones from *Polygonum cuspidatum* as tyrosynase inhibitors for dermal use. *Phytother. Res.*, 22:552–556.

Li, H., C. Zhou, Y. Pan, X. Gao, X. Wu, H. Bai, L. Zhou, Z. Chen, S. Zhang, S. Shi, J. Luo, J. Xu, L. Chen, X. Zheng, and Y. Zhao. 2005. Evaluation of antiviral activity of compounds isolated from *Ranunculus sieboldii* and *Ranunculus sceleratus*. *Planta Med.*, 71:1128–1133.

Li, L., G.E. Henry, and N.P. Seeram. 2009. Identification and bioactivities of resveratrol oligomers and flavonoids from *Carex folliculata* seeds. *J. Agric. Food Chem.*, 57:7282–7287.

Li, L., and W.H. Hou. 2007. Inhibitory effects of liquor cultured with *Nelumbo nucifera* and *Nymphaea tetragona* on the growth of *Microcystis aeruginosa*. *Huan Jing Ka Xue*, 28:2180–2186.

Li, T.S.C. 2011. *Taiwanese native medicinal plants—Phytopharmacology and therapeutic values*. CRC Taylor & Francis, New York.

Liang, D., Z.Y. Hao, G.J. Zhang, Q.J. Zhang, R.Y. Chen, and D.Q. Yu. 2011. Cytotoxic triterpenoid saponins from *Lysimachia clethroides*. *J. Nat. Prod.*, 74:2128–2136.

Liao, Y.J., H.F. Zhai, B. Zhang, T.X. Duan, and J.M. Huang. 2011. Anxiolytic and sedative effects of dehydroeffusol from *Juncus effuses* in mice. *Planta Med.*, B77:416–420.

Lin, F., L. Chen, E. Lin, F. Ke, H. Chen, M. Tsai, and P. Hsiao. 2007. Compounds from *Wedelia chinensis* synergistically suppress androgen activity and growth in prostate cancer cells. *Carcinogenesis*, 28:2521–2529.

Lin, H.W., M.X. Sun, Y.H. Wang, L.M. Yang, Y.R. Yang, N. Huang, L.J. Xuan, Y.M. Xu, D.L. Bai, Y.T. Zheng, and K. Xiao. 2010. Anti-HIV activities of the compounds isolated from *Polygonum curpidatum* and *Polygonum multiflorum*. *Planta Med.*, 76:889–892.

Liu, X.T., Q.S.Y. Pan, I.D. Williams, H.H. Sung, Q. Zhang, J.Y. Liang, N.Y. Ip, and Z.D. Min. 2006. Ent-rosane and labdane diterpenoids from *Sagittaria sagittifolia* and their antibacterial activity against three oral pathogens. *J. Nat. Prod.*, 69:255–260.

Londonkar, R.L., and P.V. Poddar. 2009. Studies on activity of various extracts of *Mentha arvensis* Linn. against drug induced gastric ulcer in mammals. *World J. Gastrointest. Oncol.*, 1:82–88.

Lugo, M.T.G., Y. Wang, S.G. Franzblan, E. Suarez, and B.N. Timmermann. 2005. Antitubercular sterols from *Thalia multiflora* Horkel ex Koernicke. *Phytother Res.*, 19:876–880.

Mallick, N., S. Biswas, M.C. Varma, and P.P. Pankaj. 2010. In vitro antimalarial activity of c-phycocyanin from *Nostoc muscorum. The Biscon* (Special Issue) 1:69–78.

Mandana, B., A.R. Russly, G. Ali, and S.T. Farah. 2011. Antioxidant activity of spearmint (*Mentha spicata* L.) leaves extracts by supercritical carbon-dioxide (SC-CO_2) extraction. *Int. Food Res. J.*, 18:543–547.

Marzonk, M.S., F.M. Soliman, I.A. Shehata, M. Rabee, and G.A. Fawzy. 2007. Flavonoids and biological activities of *Jussiaea repens. Nat. Prod. Res.*, 21:436–443.

Masoko, P., M.P. Mokgotho, V.G. Mbazima, and L.J. Mampuru. 2008. Biological activities of *Typha capensis* (Typhaceae) from Limpopo province (South Africa). *Afr. J. Biotechnol.*, 7:3743–3748.

Maucourt, K., M. Agarwal, B. Rene, and S. Fermandjian. 2002. Use of *Chlamydomonas reinhardtii* mutants for anticancer drug screening. *Biochem. Pharmacol.*, 64:1125–1131.

Mazid, M.A., L. Nahar, B.K. Datta, S.A.M.K. Bashar, and S.D. Sarkar. 2011. Potential antitumor activity of two *Polygonum* species. *Arch. Biol. Sci. Belgrade*, 63:465–468.

McClatchey, W.C., G.B. Mahady, B.C. Bennett, L. Shiels, and V. Salvo. 2009. Ethnobotany as a pharmacological research tool and recent developments in CNS- active natural products from ethnobotanical sources. *Pharmacol. Ther.*, 123:239–254.

Meira, M., E.P. de Silva, J.M. David, and J.P. David. 2012. Review of the genus *Ipomoea*: Traditional uses, chemistry and biological activities. *Rev. Bras. Farmacogn.*, 22(3):682–713.

Meizoso, I.R., L. Jaime, S. Santovo, F.J. Senorans, A. Cifuentes, and E. Ibanez. 2010. Subcritical water extraction and characterization of bioactive compounds from *Haematococcus fluvialis. J. Pharm. Biomed. Anal.*, 51:456–463.

Midori, T., K. Takao, K. Satoshi, T. Harukuni, and N. Hoyoku. 2001. Cancer chemopreventive activity of phenylpropanoid esters of surose, vanicoside B and lapathoside A from *Polygonum lepathifolium. Cancer Lett.*, 173:133–138.

Mishra, P.K., R. Shukla, P. Singh, B. Prakash, and N.K. Dubey. 2012. Antifungal and antiaflatoxigenic efficacy of *Caesulia axillaris* Roxb. essential oil against fungi deteriorating some herbal raw materials and its antioxidant activity. *Indust. Cr. Prod.*, 36:74–80.

Mitscher, L.A., Y.H. Park, A. Alshamma, P.B. Hudson, and T. Haas. 1981. Amorfrutin A and B, bibenzyl antimicrobial agents from *Amorpha fruticosa. Phytochemistry*, 20:781–785.

Mo, S., A. Krunic, G. Chlipala, and J. Orjala. 2009. Antimicrobial ambiguine isonitriles from the cyanobacterium *Fischerella ambigua. J. Nat. Prod.* 5:894–899.

Mohan Muruga Raja, M.K., N.K. Sethiya, and S.H. Mishra. 2010. A comprehensive review on *Nymphaea stellata*: A traditionally used bitter. *J. Adv. Pharm. Technol. Res.*, 1:311–319.

Moore, R.E., X.G. Yang, and G.M.L. Patterson. 1987. Fontanamide and anhydrohapaloxindole A, two new alkaloids from the blue green algae *Hapalosiphon fontinalis. J. Org. Chem.*, 52:3773–3777.

Mukhtar, M.D., A. Sani, and A.A. Yakasai. 2004. Cytotoxicity of fractions of *Pistia stratiotes* L. on larvae of *Culex* mosquito and *A. salina. Anim. Res. Int.*, 1:95–99.

Mundt, S., S. Kreitlon, and R. Jansen. 2003. Fatty acids with antibacterial activity from the cyanobacterium *Oscillatoria redekei* HUB 051. *J. Appl. Phycol.*, 15:263–267.

Murakami, M., S. Kodani, K. Ishida, H. Matsuda, and K. Yamaguchi. 1997. Micropeptin 103, a chymotrypsin inhibitor from the cyanobacterium *Microcystis viridis* (NIES-103). *Tetrahedron Lett.*, 38:3035–3038.

Murali, A., C. Sudha, V. Madhavan, and S.N. Yoganarasimhan. 2007. Anticonvulsant and sedative activity of Tagara (*Nymphoides macrospernum*). *Pharm. Biol.*, 45:407–410.

Nagaveni, V., and T.R. Rao. 2010. Evaluation of antioxidant potential and qualitative analysis of major polyphenols by RP-HPLC in *Nymphaea nouchali* Brum. flowers. *Int. J. Pharm. Pharmacol. Sci.*, 2(Suppl. 4):98–104.

Nakamura, Y., A. Murakami, K. Koshimizu, and H. Ohigashi. 1996. Identification of pheophorbide *a* and its related compounds as possible antitumor promoters in the leaves of *Neptunia oleracea*. *Biosci. Biotechnol. Biochem.*, 60:1028–1030.

Namikoshi, M., K.L. Rinehart, R. Sakai, R.R. Stotts, A.M. Dahlem, V.R. Beasley, W.W. Carmichael, and W.R. Evans. 1992. Identification of 12 hepatotoxins from a Homer Lake bloom of the cyanobacteria, *Microcystis aeruginosa*, *Microcystis viridis* and *Microcystis wesenbergii*: Nine new microcystins. *J. Org. Chem.*, 57:866–872.

Navin, C., and P.K. Rohatagi. 2005. Impact toughness of *Ipomaea carnea* particulate polyester composite. *J. Mat. Sci. Lett.*, 6:695–697.

Niesen, D., A.G. Sarrias, H. Ma, T. Yuan, G.E. Henry, and N.P. Seeram. 2012. Isolation of cytotoxic constituents from *Carex vulpinoidea*. *Planta Med.*, 78:1306.

Nikolova, M., L. Evstatieva, and T.D. Nguyen. 2011. Screening of plant extracts for antioxidant properties. *Botanica Serbica*, 35:43–48.

Noaman, N.H., A. Fattah, M. Khaleafa, and S.H. Zaky. 2004. Factors affecting antimicrobial activity of *Synechococcus neopoliensis*. *Microbiol. Res.*, 159:395–402.

Onofrejova, L., J. Vasickova, B. Klejdus, P. Stratil, L. Misurcova, S. Kracmar, J. Kopecky, and J. Vacek. 2010. Bioactive phenols in algae: The application of pressurized liquid and solid-phase extraction techniques. *J. Pharm. Biomed. Anal.*, 51:464–470.

Orhan, I., B. Sener, T. Atici, R. Brun, R. Perozzo, and D. Tasdemir. 2006. Turkish freshwater and marine macrophyte extracts show in vitro antiprotozoal activity and inhibit Fabl, a key enzyme of *Plasmodium falciparum* fatty acid biosynthesis. *Phytomedicine*, 13:388–393.

Ostensvik, O., O.M. Skulberg, B. Underdal, and V. Hormazabal. 1998. Antibacterial properties of extracts from selected planktonic freshwater cyanobacteria. A comparative study of bacterial assays. *J. Appl. Microbiol.*, 84:1117–1124.

Oyedeji, O., M. Oziegbe, and F.O. Taiwo. 2011. Antibacterial, antifungal and phytochemical analysis of crude extracts from the leaves of *Ludwigia abyssinia* A. Rich and *Ludwigia decurrens* Walter. *J. Med. Pl. Res.*, 5:1192–1199.

Ozer, J., N. Eisner, E. Ostrozhenkova, A. Bacher, W. Eisenreich, D. Benharroch, A.G. Goldhirsh, and J. Gopas. 2009. *Nuphar lutea* thioalkaloids inhibit the nuclear factor kappa B pathway, potentiate apoptosis and are synergic with cisplatin and etoposide. *Cancer Biol. Ther.*, 8:1860–1868.

Pal, D.K., A.K. Padhihari, M. Otta, S. Khatun, S. Sanigrahi, and M. Mandal. 2004. Studies on antibacterial activity of *Hydrilla verticillata*. In *16th Annual Conference of PSI*, Paschim, Medinipur, p. 77.

Parek, J., and S. Chanda. 2007. In vitro antimicrobial activity of *Trapa natans* L. fruit rind extracted in different solvents. *Afr. J. Biotechnol.*, 6:766–770.

Park, Y.S., K. Towantakavanit, T. Kowalska, S.T. Jung, K.S. Ham, B.G. Heo, J.Y. Cho, J.G. Yun, H.J. Kim, and S. Gorinstein. 2009. Bioactive compounds and antioxidant and antiproliferative activities of Korean white lotus cultivars. *J. Med. Food*, 12:1057–1064.

Patel, S., D. Banji, O.J.F. Banji, M.M. Patel, and K.K. Sha. 2010. Scrutinizing the role of aqueous extract of *Trapa bispinosa* as an immunomodulator in experimental animals. *Int. J. Res. Pharm. Sci.*, 1:13–19.

Patel, V., S. Shukla, and S. Patel. 2009. Free radical scavenging activity of *Grangea maderaspatana* Poir. *Pharmacog. Mag.*, 5:381–387.

Patil, S., S. Narayanan, G. Eibl, and C.I. Jolly. 2004. Evaluation of antimitotic activity of *Rotula aguatica* (Lour) a traditional herb used in treatment of cancer. *Indian J. Exp. Biol.*, 42:893–899.

Pervin, F., M.M. Hossain, S. Khatun, S.P. Siddique, K.A. Salam, M.R. Karim, and N. Absar. 2006. Comparative cytotoxity study of six bioactive lectins purified from pond weed, *Potamogeton nodosus* root stock on brine shrimp. *J. Med. Sci.*, 6:999–1002.

Pittella, F., R.C. Dutra, D.D. Junior, M.T.P. Lopes, and N.R. Barbosa. 2009. Antioxidant and cytotoxic activities of *Centella asiatica* (L) Urb. *Int. J. Mol. Sci.*, 10:3713–3721.

Podolak, I., Z. Janeczko, A. Galanty, M. Michalik, and D. Trojanowska. 2007. A triterpene saponin from *Lysimachia thrysiflora* L. *Acta Pol. Pharm. Drug Res.*, 64:39–43.

Prabu, K., N. Kanchana, and A.M. Sadiq. 2011. Hepatoprotective effect of *Eclipta alba* on paracetamol-induced liver toxicity in rats. *J. Microbiol. Biotechnol. Res.*, 1:75–79.

Prakash, J.W., J. Marimuthu, and S. Jeeva. 2011. Antimicrobial activity of certain freshwater microalgae from Thamirabarani River, Tamilnadu, South India. *Asian Pac. J. Trop. Biomed.*, S170–S173.

Qadar, S.W., M.A. Abdulla, L.S. Chua, and S. Handan. 2012. Potential bioactive property of *Polygonum minus* Huds (Kesum) review. *Sci. Res. Essays*, 7:90–93.

Rahman, A., A. Sayeed, A. Islam, D. Chowdhury, G. Sadik, and G.R.M.A.M. Khan. 2002. Characterization and biological screening of a triterpenoid from *Nymphoides cristatum*. *J. Biol. Sci.*, 2:46–48.

Rajeshwari, K.R., and M. Rajashekhar. 2011. Biochemical composition of seven species of cyanobacteria isolated from different aquatic habitats of Western Ghats of Southern India. *Braz. Arch. Biol. Technol.*, 54.

Ramamurthy, V., S. Raveendran, S. Thirumeni, and S. Krishnaveni. 2012. Antimicrobial activity of heterocytic cyanobacteria. *Int. J. Adv. Life Sci.*, 1:32–38.

Rao, A.S., and A. Merugu. 2012. Shikimic acid as a major compound from *Ludwigia alternifolia* Linn. *J. Pharmacog.* 3:121–122.

Rao, P.V.L. 2004. Cyanobacterial toxins: Effects and control measure. In *Pharmacological perspectives of toxic chemicals and their antidotes*, ed. S.J.S. Flora, J.A. Romano, S.I. Baskin, and K. Sekhar, 179–192. Narosa Publishing House, New Delhi, India.

Rao, V.M., G.L.V. Damu, D. Sudhakar, and C.V. Rao. 2009. Two new bioactive flavones from *Grangea maderaspatana* (*Artemsia maderaspatana*). *Asian J. Chem.*, 21:1552–1558.

Rasool, M., E.P. Sabina, P. Nithya, and K. Lavanya. 2008. Studies on the analgesic, antipyretic and ulcerogenic properties of *Spirulina fusiformis* in mice. *J. Pharmacol. Toxicol.*, 3:47–52.

Razik, A.F.A., A.S.I. Elshamy, M.I. Nassar, S.M.E. Kousy, and H. Hamdy. 2009. Chemical constituents and hepatoprotective activity of *Juncus subulatus*. *Rev. Latinoam. Quim.*, 37:70–84.

Ripa, F.A., L. Nahar, M. Haque, and M.M. Islam. 2009. Antibacterial, cytotoxic and antioxidant activity of crude extract of *Marsilea quadifolia*. *Eur. J. Sci. Res.*, 33:123–129.

Row, L., C. Chan, and J. Ho, 2004. One alkenylphenol and steroids from the aquatic plant *Monochoria vaginalis*. *J. Chin. Chem. Soc.*, 51:225–228.

Saeed, A. 2003. Synthesis and bioactivity of xyridin A and B metabolites from *Xyris indica*. *J. Heterocyclic Chem.*, 40:337–340.

Sannigrahi, S., U.K. Mazumder, A. Mondal, D. Pal, S.L. Mishra, and S. Roy. 2010. Flavonoids of *Enhydra fluctuans* exhibit anticancer activity against Ehrlich's ascites carcinoma in mice. *Nat. Prod. Commun.*, 5:1239–1242.

Sannigrahi, S., U.K. Mazumder, D. Pal, M.L. Mishra, and S. Maity. 2011. Flavonoids of *Enhydrina fluctuans* exhibit analgesic and anti-inflammatory activity in different animal models. *Pakistan J. Pharm. Sci.*, 24:369–375.

Sano, T., J. He, Y. Lin, and K. Kaya. 1998. Isolation of bioactive compounds in cyanobacteria from Chinese freshwater. 1. Trypsin inhibitor. *Phycol. Res.*, 46(Suppl. 2):13–17.

Santos, N.D., D.L. Gomes, L.S. Costa, S.L. Cordeiro, M.S.S.P. Costa, E.S. Trinidade, C.R.C. Franco, K.C. Scortecci, E.L. Leite, and H.A.O. Rocha. 2012. Freshwater plants synthesize sulfated polysaccharides: Heterogalactans from water hyacinth (*Eicchornia crassipes*). *Int. J. Mol. Sci.*, 13:961–976.

Selvakumari, E., S. Shantha, T.P. Prabhu, and C. Sreenathkumar. 2012. Antiproliferative activity of ethanolic flower extract from *Nymphaea pubescens* Willd against human cervical and breast carcinoma in vitro. *Int. Res. J. Pharm.*, 3:124–125.

Shanab, S.M.M. 2007. Bioactive allele-chemical compounds from *Oscillatoria* species (Egyptian isolates). *Int. J. Agric. Biol.*, 9:617–621.

Shanab, S.M.M., E.A. Shalaby, D.A. Lightfoot, and H.A. El-Shemy. 2010. Allelopathic effects of water hyacinth *Eichhornia crassipes. PLoS One*, 5:e13200.

Sharathchandra, K., and M. Rajashekhar. 2011. Total lipid and fatty acid composition in some freshwater cyanobacteria. *J. Algal Biomass Utln.*, 2:83–97.

Shirwalkar, A., K.S. Prabhu, and I.S.R. Punitha. 2006. In vitro antioxidant studies of *Sphaeranthus indicus* (Linn.). *Indian J. Exp. Biol.*, 44:993–996.

Shokravi, S., G. Zarrini, Y. Ghasemi, M.T. Yazdi, A. Shafice, and M. Armini. 2004. Parsiguine, a novel antimicrobial substance from *Fischerella ambigua. Pharm. Biol.*, 42:318–322.

Shyur, L., J.H. Tsung, J. Chen, C.Y. Chiu, and C.P. Lo. 2005. Antioxidant properties of extracts from medicinal plants popularly used in Taiwan. *Int. J. Appl. Sci. Eng.*, 3:195–202.

Singh, A.P., and B.R. Chaudhary. 2010. Preliminary phycochemical analysis and in vitro antibacterial screening of *Pithophora oedogonia* (Mont.) Wiltrock—A fresh water green algae forming mats in the water bodies *J. Algal Biomass Utln.*, 1:33–41.

Singh, S., K.N. Bhusan, and V.C. Banerjee. 2005. Orbit—Red your bioactive compounds from cyanobacteria and microalgae—An overview (online). *Red Orbit Universe*, October 23.

Sivarajan, V.V., and I. Balachandran. 1994. *Ayurvedic drugs and their plant sources*, 358–359. Oxford & IBH Publishing Co., New Delhi, India.

Small E., and P.M. Catling. 1999. *Canadian medicinal crops*. National Research Council, Canada, NRC Research Press.

Smith, J.L., G.L. Boyer, and P.V. Zimba. 2008. A review of cyanobacterial odorous and bioactive metabolites: Impacts and management alternatives in aquaculture. *Aquaculture*, 280:5–20.

Soltani.S., S. Saadartmand, R. Khavarinejad, and T. Nejadsattri. 2011. Antioxidant and antibacterial activities of *Cladophora glomerata* (L) Kutz, in Caspian Sea Coast, Iran. *Afr. J. Biotechnol*, 10:7684–7689.

Song, C.W., S.M. Wang, L.L. Zhou, F.F. Hou, K.J. Wang, Q.B. Han, N. Li, and Y.X. Cheng. 2011. Isolation and identification of components responsible for antioxidant capacity of *Euryale ferox* seeds. *J. Agric. Food Chem.*, 59:1199–1204.

Sowemimo, A., M. van de Venter, L. Baatjies, and T. Koekemoer. 2009. Cytotoxic activity of selected Nigerian plants. *Afr. J. Trad. Complement Altern. Med.*, 6:526–528.

Srivastava, V.C. 1996. Inhibitory metabolite production by the cyanobacterium *Fischerella musicola*. PhD thesis, Biotechnology and Bioprocess Engineering, Massey University.

Staba, E., S. John, and K. Lee. 1972. *Aquatic plants from Minnesota. Part 2. Toxicity, antineoplastic and coagulant effects*. Water Resources Research Centre (WRRC) Bulletin 47.

Sultanova, N., T. Makhmoor, Z.A. Abilov, Z. Parween, V.B. Omur Kamzinova, A. ur-Rahman, and M.I. Choudhary. 2001. Antioxidant and antimicrobial activities of *Tamarix ramosissima. J. Ethnopharmacol.*, 78:201–205.

Swapna, M.M., R. Prakashkumar, K.P. Anoop, C.N. Manju, and N.P. Ranjith. 2011. A review on the medicinal and edible aspects of aquatic and wetland plants of India. *J. Med. Plant Res.*, 5:7163–7176.

Tellez, M.R., K.K. Sehrader, and M. Kobaisy. 2001. Volatile components of the cyanobacterium *Oscillatoria perronata* (Skuja). *J. Agric. Food Chem.*, 49:5989–5992.

Thiamdao, S., M. Motham, J. Pekkoh, L. Mungmai, and Y. Peerapornpisal. 2012. *Nostochopsis lobatus* wood, em. Geitler (Nostocales), edible algae in northern Thailand. *Chiang Mai J. Sci.*, 39(1):119–127.

Thippeswamy, B.S., B. Mishra, V.P. Veerapur, and G. Gupta. 2011. Anxiolytic activity of *Nymphaea alba* Linn. in mice as experimental models of anxiety. *Indian J. Pharmacol.*, 43:50–55.

Trimurtulu, G., I. Ohtani, G.M.L. Patterson, R.E. Moore, T.H. Corbett, F.A. Valeriote, and L. Demchik. 1994. Total structures of cryptophycins, potent antitumor depsipeptides from the blue-green alga *Nostoc* sp. strain GSV 224. *J. Am. Chem. Soc.*, 116:4729–4737.

Uddin, M.E., A.M.T. Islam, M.A. Chowdhury, M.K. Rahman, M.S. Islam, and M.R. Islam. 2012. Sedative and analgesic activities of *Ludwigia repens. Pharmacology*, 2:202–211.

Victory, K.J., D.M. Lewis, and J.D. Iddle. 2010. Identification of a pharmacologically active peptide of *Microcysti flos-aquae*. Archivos/labcontrol.Cl/wcce 8/offline/techsched/mss/60uxlu.pdf.

Wang, Y.G., Y.L. Wang, H.F. Zhai, Y.J. Liao, B. Zhang, and J.M. Huang. 2012. Phenanthrenes from *Juncus effusus* anti anxiolytic and sedative activities. *Nat. Prod. Res.*, 26:1234–1239.

Wangchuk, P. 2004. Bioactive alkaloids from medicinal plants of Bhutan. MSc thesis, University of Wollongong, Australia.

Wojdylo, A., J. Oszmianski, and R. Czemerys. 2007. Antioxidant activity and phenolic compounds in 32 selected herbs. *Food Chem.*, 105:940–949.

Xian, Q., H. Chen, H. Liu, H. Zou, and D. Yin. 2006. Isolation and identification of antialgal compounds from the leaves of *Vallisneria spiralis* L. by activity guided fractionation. *Environ. Sci. Pollut. Res. Int.*, 13:233–237.

Yakob, H.K., S.F. Sulaiman, and A.M. Uyub. 2012. Antioxidant and antibacterial activity of *Ludwigia octovalis* on *Escherichia coli* 0157:H7 and some pathogenic bacteria. *World Appl. Sci. J.*, 16:22–29.

Yang, S.L., L.H. Lihua Tang, J.K. Tian, J. Guo, C. Xie, Q.M. Xu, and L.Z. Xu. 2009. Antitumor constituents of four medicinal plants from *Lysimachia* genus. *Plant Med.*, 75(4).

Yoshikawa, M., S. Yoshizumi, T. Murakami, H. Matsuda, J. Yamahara, and N. Murakami. 1996. Medicinal food stuffs. II. On the bioactive constituents of the tuber of *Sagittaria trifolia* L. (Kuwai, Alismataceae): Absolute stereostructures of trifoliones A, B, C, and D, sagittariosides *a* and *b*, and arabinothalictoside. *Chem. Pharm. Bull.* (Tokyo), 44:492–499.

Ysrael, M.C. 2003. Tankin herbal drug: A multidisciplinary approach to development. *Clin. Hemorheol. Microcirc.*, 29:247–251.

Zelenski, S.G. 1997. Alkaloids of *Nelumbo lutea* (Nymphaceae) *J. Pharm. Sci.*, 66:627–628.

Zhang, Z., H.N. Elsohly, X.C. Li, S.I. Khan, S.E. Broedel Jr., R.E. Raulli, R.C. Cihlar, C. Burandt, and L.A. Walker. 2003. Phenolic compounds from *Nymphaea odorata. J. Nat. Prod.*, 66:548–550.

Zhao, J., T. Liu, L. Ma, M. Yan, Z. Gu, Y. Huang, F. Xu, and Y. Zhao. 2011. Antioxidant and preventive effects of extract from *Nymphaea candida* flower in in vitro immunological liver injury of rat primary hepatocyte cultures. *Evid. Based Complement Alternat. Med.*, 2011:497–673.

Compounds

C

D

E

Index

A

B

C

R

S